Reverse Engineering Social Media

Reverse Engineering Social Media

Software, Culture, and Political Economy in New Media Capitalism

Robert W. Gehl

TEMPLE UNIVERSITY PRESS
PHILADELPHIA

TEMPLE UNIVERSITY PRESS
Philadelphia, Pennsylvania 19122
www.temple.edu/tempress

Copyright © 2014 by Temple University
All rights reserved
Published 2014

Library of Congress Cataloging-in-Publication Data

Gehl, Robert W.
  Reverse engineering social media : software, culture, and political
economy in new media capitalism / Robert W. Gehl.
      pages cm
  Includes bibliographical references and index.
  ISBN 978-1-4399-1034-4 (hardback : alk. paper) — ISBN 978-1-4399-1035-1
(paper : alk. paper) — ISBN 978-1-4399-1036-8 (e-book)   1. Online social
networks.   2. Social media.   I. Title.
  HM742.G44 2014
  302.3—dc23
                                                          2013042506

♾ The paper used in this publication meets the requirements of the American
National Standard for Information Sciences—Permanence of Paper for Printed
Library Materials, ANSI Z39.48-1992

Printed in the United States of America

2 4 6 8 9 7 5 3 1

To all those who are fighting for a better media system,
working uphill to build alternatives to mainstream social media,
exposing the internal details of the surveillance system,
or contributing to free software

Contents

Acknowledgments

Anyone who takes seriously actor-network theory knows that a book is a punctualization of a large, heterogeneous network of many ingredients. Nonhuman actors who contributed to this book include my dogs, my pet praying mantis, and the mountains outside my window—not to mention Treasure Valley Coffee, Linux Mint (with Mate, of course), Zotero, Gnome Do, Libreoffice, computer screens, nineties grunge rock and seventies soul, hundreds of books, guitars, wireless mice, and science fiction movies.

But I will spend more time thanking humans. First, I thank the staff and editors at Temple University Press. Mick Gusinde-Duffy got the ball rolling, and Micah Kleit picked it up and finished the job. I hope they are as happy with this work as I have been working with them. I especially want to thank Heather Wilcox and Joan Vidal, who have put this book through the copyediting wringer; it's a far better book for it!

A special thank-you goes to the Tanner Humanities Center at the University of Utah, especially to Bob Goldberg, Beth Tracy, and Josh Elstein. The Tanner Center hosted me as an Aldrich Fellow in the spring of 2013, which allowed me to work on this book without any distractions. In addition, my fellow Fellows at the center helped workshop ideas with me. Many of their suggestions are reflected in the text.

In addition, I owe a great deal to my colleagues in what I jokingly call the "Will THIS Get Me Tenure?" Writing Group at the University of Utah: Anya Plutynski, Mike Middleton, Casey Boyle, and Danielle Endres. Mike and Casey, especially, sparked ideas that appear in this book. Mike made very important suggestions early in the life of Chapter 1. And Danielle has been the heart of the writing group, showing us all what it means to write often and write well.

At Utah, I have enjoyed the support and intellectual camaraderie of all my colleagues in the Department of Communication, especially Mary Strine, Suhi Choi, Len Hawes, Jake Jensen, Bob Avery, Ann Darling, Heather Canary, Robin Jensen, Helga Shugart, the Jims (Jim Anderson and Jim Fisher, roof builders extraordinaire), Marouf Hasian, Joy Pierce, Kevin Deluca, Ye Sun, and Glen Feighery. Special thanks go to Connie Bullis, who helped expand my perspective on the field of communication and gave me the confidence to work with the graduate students; Avery Holton and Julianna Holton, who were happy and willing babysitters; Kevin Coe and Julia Coe, who watched my pet Chinese praying mantis; and Kent Ono and family for their support and guidance.

I also have students who contributed to this project by, well, being great students and pushing my thinking, especially Amanda Friz-Siska, Sarah Bell, and the entire Culture of Computing undergraduate course. I was so impressed by the students in that class, and they had such a great impact on the writing of this book, that I told them I would thank them by name. As promised, here goes: Jace Bradford, Tyler Pratt, Colin Cronin, Kaylynne Hatch, Cramer England, Chet Cannon, Jeff Dunn, John Foote, Spencer Broste, Katy White, Lucy Shephard, Andreas Riviera, Lauren Doxey, and Kyle Biehl. What a great class!

Much of who I am intellectually has come from my friends and mentors at George Mason University's Cultural Studies program. I benefited from working with Paul Smith and Roger Lancaster. I have too many friends among the students to name them all, but the ones who had a direct impact on this work are Jarrod "Ragin'" Waetjen, Vicki Watts (5G sends love to Adelaide!), Randall Cohn, Randa Kayyali,

Fan Yang (as predicted, you won the TCS race!), Kristin Scott, Lia Uy-Tioco, and Tara Sheoran. I especially thank my dissertation committee: Tim Gibson, Alison Landsberg, Mark Sample, and—above all—Hugh Gusterson. If I can be half the teacher Hugh is—and if I can be half as intellectually brave as he is—I will be successful.

A great deal of this work was presented at conferences far and wide, and I owe much to my colleagues and friends in several conference circuits. The Cultural Studies Association has become an intellectual home to me, and I have had great discussions about theory and culture with Jaafar Aksikas, Sean Andrews, Jamie "Skye" Bianco, Meg Turner, Bruce Burgett, Ted Striphas, and Patricia Clough. I have also had the privilege of being invited to present at Compromised Data, hosted by Greg Elmer, Ganaele Langlois, and Joanna Redden at Ryerson University and attended by such great scholars as Jean Burgess, Carolyn Gerlitz, Taina Bucher, Axel Bruns, Alessandra Renzi, and Fenwick McKelvey. It was as though one of my Works Cited pages came to life and drank beer with me. Finally, and probably most importantly, I was invited to present at the first Unlike Us conference in Cyprus. There I met Geert Lovink and Korinna Patelis, and I got my first up-close exposure to the world of social media alternatives (such as Mark Stumpel's wonderful Facebook Resistance project). Since then, Lovink and Patelis have put on multiple Unlike Us events, bringing together critical scholars and hacktivists to think through what it means to build a better social media system. This book is tremendously influenced by those conferences.

Portions of this work appeared in several journals. Material from the Introduction was originally published as R. W. Gehl and Sarah Bell, "Heterogeneous Software Engineering: Garmisch 1968, Microsoft Vista, and a Methodology for Software Studies," *Computational Culture*, no. 2 (2012). Material from Chapter 1 was originally published as Robert Gehl, "What's on Your Mind? Social Media Monopolies and Noopwer," *First Monday* 18, no. 3/4 (March 2013). An earlier version of Chapter 2 was originally published as Robert Gehl, "The Archive and the Processor: The Internal Logic of Web 2.0," *New Media and Society* 13, no. 8 (December 2011): 1228–1244. An earlier version of Chapter 3 was originally published as Robert Gehl, "Real (Software) Abstractions:

On the Rise of Facebook and the Fall of Myspace," *Social Text* 30, no. 2 111 (2012): 99–119 (copyright 2012, Duke University Press; all rights reserved; reprinted by permission of the present publisher, Duke University Press; www.dukeupress.edu). I am grateful for the feedback I received from the anonymous reviewers and editors involved with these early publications, especially Matt Fuller and Jonathan Beller.

Finally, closer to home, I thank my new friends Sean Lawson and Cynthia Love, who are the best people in the world to eat ribs and watch football with. I am grateful to my oldest friends, Ry, Mony, Brian, and Dan; every summer, they remind me how to kick it old school. I thank my in-laws, especially Captain Joe, for taking me fishing and camping in the summers to help me get away from work. I gratefully acknowledge my mom and dad and brother; I cannot begin to trace how their love has shaped my work. And I thank the two TJs: Teddy, my son, who is right now refusing to nap because he is a happy guy who loves to play, and Jesse Houf, my partner, who more than anyone in the world remains my inspiration to keep working and who is most responsible for my earning a Ph.D.

Thank you all.

Introduction
Looking Forward and Backward

Heterogeneous Engineering
of Social Media Software

sumoto.iki's web2diZZaster

Web artist sumoto.iki's "web2diZZaster" is a collection of bland, muted pastel images containing little more than rectangles and lines.[1] The images are unremarkable, even unattractive, and it is hard to determine what they represent. And yet, many of these images seem eerily familiar. A second glance reveals why: these muted rectangles take shape as common social media sites. Digg, the social bookmarking site, is identifiable by the peach tabs that indicate the number of "diggs" that users have given to various stories. The tabs are empty, as is the rest of the page, but this largely empty frame is still recognizable. Myspace, the failed (and resurrected) social-networking site, is identifiable by its blue banner and log-in fields, and directly below it is Facebook's even more sedate (but much more successful) home page. YouTube is harder to recognize until the viewer sees the iconic red polygon and the two series of four rectangles where featured videos normally appear. Although its design has changed since sumoto. iki made this project, Twitter is perhaps the most recognizable because of its light blue field and narrow, prominent center column.

I am having trouble describing the momentarily unsettled response I had to sumoto.iki's art. However, after a moment of squinting at the images and contemplating, I realized that sumoto.iki presents all these flagship social media sites *without any user content*. Without comments, videos, lists of friends, and editorial recommendations, these sites appear as "ghostly forms," which sumoto.iki describes as a "first impression of a possible apocalypse where only HTML vestiges would remain inside a dehumanized network of all networks." The disaster in "web2-diZZaster" is the social Web at the end of humanity. It is the tragic silence of a world without tweets, status updates, YouTube videos, diggs, and Facebook connections. When I viewed this art, all I had to go on were the surface manifestations of the most basic elements of HTML, PHP: Hypertext Preprocessor, Javascript, and CSS: div tags, positions and floats, colors, and lifeless scripts. Within the "dehumanized network" and without user-generated content, these social media sites appear as mere frames, and unappealing ones at that. Without content, these sites are lifeless shells. Without it, social media cannot work.

Of course, social media are working just fine precisely because users *do* contribute so much to these frames. Nearly a billion people populate the social network Facebook, creating constant streams of comments, links, "likes," and applications. Twitter's meteoric growth is also measured in user-generated content. Social bookmarking and link-sharing sites Digg and Reddit command millions of page views, and an upvoted link posted on their pages can drive traffic to websites. Amazon posts millions of user-written reviews of books and products. LinkedIn is filled with job-seekers and headhunters posting résumés and e-introducing one another. Flickr has billions of photos and comments, and Facebook's servers contain the largest collection of user-uploaded photographs in the world thanks to its ownership of Instagram. And the company that is perhaps the exemplar of social media is Google, which relies on user-generated links, videos, social connections, and blogs to power its highly profitable search and advertising business. Investors who participated in the 2004 Google IPO have seen their investment grow substantially; the stock was offered at $85 and as of this writing trades for nearly $1,100. Although Facebook's IPO was much less successful, it still generated $1 billion. Twitter's recent IPO avoided Facebook's missteps to raise nearly $2 billion. This rise in value comes directly from user-created content, all within the opposite of sumoto.iki's "dehumanized" network. Perhaps we should borrow a phrase from networking company Cisco and call it the "Human Network."

Considering the history of the Web, it seems unlikely that social media would be such a commercial success. Just over a decade ago, the term "dot-com" (i.e., commerce on the Web) drew derision from anyone with an interest in business. The 2000–2001 financial/technology bubble burst resulted from irrational exuberance; investors in online commercial sites, such as Pets.com, lost millions of dollars when that business model failed to catch on. Direct-to-consumer sales of pet supplies, groceries, and gardening supplies had all the sustainability of a paper fire. In addition, investment in the networks themselves—that is, in the physical connections between sites—was drawn to oversaturated corridors, such as between New York and Boston. Just like the housing bubble burst of 2008, in the late 1990s there was a fiber-optic bubble.[2] When it burst, investors withdrew from the market almost immediately. According to PricewaterhouseCoopers, the first quarter of 2000 saw investment in IT peak at $2.8 billion, and the first quarter of 2002 saw investment of only 10 percent of that peak level.[3] Even in the post-Google IPO years, investment has come nowhere close to even 25 percent of the peak of the bubble years.

While the 2000–2001 market rejection of direct-to-consumer marketing of mass-produced goods and online commerce scared venture capitalists seeking to profit from the Web, another more sustained contemporary movement was attacking global capitalism, mass culture, and private property—and relying on the Web to do so. Writing about the Zapatista movement of the 1990s, Maria Elena Martinez-Torres notes that "a paradox has emerged from the revolution in communications: the same technology that has taken world capitalism to a new stage of development—corporate globalization—has also provided a significant boost for anti-corporate and anti-globalization movements."[4] By co-opting the Internet as a space of spectacle and image politics,[5] antiglobalization and progressive movements have been able to transmit their messages to worldwide audiences.[6] Even in the midst of the dot-com euphoria of 1999–2000, protesters were able to use the Web to organize massive, coordinated demonstrations against the World Trade Organization (WTO) meeting in Seattle. The actions of the estimated forty thousand protesters were supplemented with the advent of Indymedia, a user-led, anticapitalist news source that began as an alternative to mainstream coverage of the Seattle protests. These anticapitalist uses of the Web were engagements in what Nick Dyer-Witheford calls the "struggle for the general intellect."[7] Drawing on Karl Marx's iconoclastic "Fragment on machines" in the *Grundrisse*,[8] Dyer-Witheford argues that the Internet has simultaneously enabled extensions of the Taylorist domination of labor and the very means for labor to short-circuit global capital. On the one hand, the Internet might allow for "fast capitalist"[9] flows of commodities and value realization, but on the other hand, it allows for the fast and space-eroding coordination of protest.

Here, we see two interweaving movements going online. On the one hand, capitalism's cycles of boom and bust came to the Web as the irrational rationality of herdlike investment movements seized on cyberspace as the next great marketplace.[10] On the other hand, the strong counterhegemonic possibilities of the Internet and Web were being explored by an increasing number of "hacktivists," cybersocialists, and antiglobalization agitators seeking to appropriate the Web as a tool of revolution.

This seems to be an unlikely place for social media capitalism to thrive. However, in 2004, technology book publisher Tim O'Reilly and journalist and technology blogger John Battelle confidently stood on stage at the inaugural "Web 2.0 Conference" in San Francisco and argued that Web-based commerce was making a comeback.[11] Their proof came from companies that recognized the chaotic, unpredictable nature of user-generated discourse on the Web and were able to create sites that harnessed this "collective intelligence." In O'Reilly and Battelle's vision of Web 2.0, companies that were appropriating the socially articulated energies, passions, and labors of users—wherever those users might go with them—were not only surviving in the world of online commerce; they were building new media empires.

In this milieu, the practices—but certainly not the anticapitalist politics—of a user-generated site such as Indymedia are replicated within the user-generated, for-profit CNN iReport, where "citizen journalists" produce and share news stories. The Zapatistas and their supporters are now welcome to connect via Facebook at the Chiapas Project.[12] The spectacle of mass protests drives news coverage, and thanks to the personalization of Google News and user-generated services Reddit and del.icio.us, we can easily keep up with the latest developments. Dyer-Witheford's manifesto *Cyber-Marx* is available on Amazon; those undecided about spending $25 on it can consider several glowing user-written reviews that appear on the listing. In short, social media are the corporate response to the mass creativity, collaboration, and desires of networked peoples. It is a tacit admission by large media companies: when given a choice, people prefer content produced and recommended by themselves and their friends to that recommended by editorial authorities. They are leery of mass culture and globalized corporations, so they seek to create their own culture.[13] They express political opinions and offer frank assessments of commodities, corporations, and states, and they openly share these opinions with their friends and colleagues.

However, it is not as though new media capitalists are simply stepping aside and allowing users to lead the way, let alone slough off capitalist media production. Clearly, social media outlets are new media capitalism's attempt to absorb and capture this explosion of user-generated content as objectified surplus value. Whatever the form—from fandom to antiglobalization manifestos—if the user-generated content occurs within what sumoto.iki calls the "ghostly frames" of many social media sites, it is being exploited for profit.

Sumoto.iki's artwork, then, is a useful illustration of the central argument of this book: social media need to be understood not just as a collection of sites that place the users at the center, nor just as a radical reform to the top-down, authoritarian model of mass media. Social media also have to be understood as software engineered to privilege and enhance certain users while closing off others. To be sure, the "ghostly frames" that sumoto.iki depicts are notable for their absence of user-generated content, but they are also notable because they *are in fact a framework for users to inhabit and populate.* This software framework contains a contradiction: on the one hand, social media allows for users to "be the media" and thus influence mass culture; on the other hand, social media sites are rigidly hierarchical, allowing certain uses and discouraging others, while site-owners constantly watch users' movements and exploit users as what Tiziana Terranova has aptly described as "free laborers."[14] As she argues, free labor is rife with contradictions: it is freely given yet exploited; it is done for love, yet hypervalorization haunts and directs it; it is work, but it is play. Similarly, the "ghostly frames" sumoto.iki reveals are simultaneously sites of user freedom and rigid structures of exploitation. Social media must be understood along these lines, and to do so, we have to turn to computer architecture design as a key source for this contradiction. As Lawrence Lessig argues, "Code is law."[15] The architecture and software matter.[16]

However, despite the limitations of and problems with the social media architecture, what do we make of social media's progressive aspects, started by the Zapatistas and the anti-WTO movement and progressing to the Arab Spring and Occupy Wall Street movements? It is myopic to only talk of exploitation while ignoring the power of new social movements. Here, I suggest that such movements indicate the pressing need for users to take an active role not only in producing online content but also in shaping the structure of the sites themselves. The fact that Occupy and the Arab Spring have achieved what they have *despite* the surveillance and exploitation of social media makes me wonder: what if social media were built to help *advance* activism and politics rather than constrain them? Again, the architecture matters. To alter social media to meet this goal, users and activists must be engaged early and often with the development of social media architecture and software to avoid many of the social inequalities and problems with privacy we currently are witnessing. Indeed, they would have to reverse engineer social media. I want this book to help point the way toward a better Web, one designed for progressive politics. In other words, I outline ways in which the structures and cultures of new media capitalism and social media software can be reverse engineered, hacked, altered, and maybe even renewed.

Software Studies and Engineering Metaphor(s)

Thus, this book is a contribution to the growing field of software studies. In the words of Rob Kitchin and Martin Dodge:

> Complementing the work of computer scientists on the mechanics of software development and human computer interaction, and research on digital technologies more generally, social theorists, media critics, and artists have begun to study the social politics of software: how it is written and developed; how software does work in the world to produce new subjects, practices, mobilities, transactions, and interactions; the nature of the software industry; and the social, economic, political, and cultural consequences of code on different domains, such as business health, education, and entertainment.[17]

Software is ubiquitous. It structures many of our spaces, mediates much of our communication, helps shape our collective and individual memories, and underpins global economics. Software is an obvious part of our daily lives as we use computers and smartphones, but it is also a hidden substrate operating out of view: running automobiles, controlling infrastructure, algorithmically calculating credit and finance, storing data on myriad human and nonhuman flows, and collecting evidence of past behaviors. It provides us with new ontologies and epistemologies as we ponder what it means to be human in an age of smart machines and smarter phones.

Software studies is a critical interrogation of this multifaceted object. Theorists and scholars working in this field study how power operates in relation to software: who writes it? Why? To what purposes? What are the conditions of its production? How does it shape uses? What is its architecture, and why was that architecture selected over competing ones? How do users reappropriate, extend, build on, or break it? What can code, layers of software, the software/hardware relationship, network topologies, and the software/user relationship tell us about our assumptions about subjectivity and identity? Software studies does more than just analyze lines of code; rather, it evaluates software as part of larger contexts, such as culture and history. Done well, a critical study of software can tell us much about the world around us.

Most histories of this new field trace it back to Lev Manovich's 2000 book *Language of New Media*, but its roots run deeper than that. Even when we limit the field to work produced after roughly 2000, there are many critical studies of software: philosophies of object-oriented programming[18] and computationalism,[19] hacking,[20] semiotic analyses of interfaces,[21] artistic reinterpretations of software (such as sumoto.iki's, described at the beginning of this chapter), cultural and technology studies,[22] legal analyses,[23] historiography,[24] spatial theorizations,[25] and critical code studies.[26] There is no single agreed-upon object or method of study, but this variety is a strength of the field; given the ubiquity of software, the field welcomes new works on new objects.

To this end, as a contribution to software studies, this book draws on three engineering metaphors, using them as methodological windows into social media software: software engineering, reverse engineering, and heterogeneous engineering.

Software Engineering

When studying software, an intuitive place to start is by running software on a machine in front of us. In the case of social media software, then, one might sign up for an account and critically consider the pedagogy and structure of the interface. I certainly have done this. However, following Matthew Kirschenbaum's warnings against "screen essentialism" and "medial ideology,"[27] I also want to extend such analyses by considering how software is made. What are its logics of production? How is the labor of designing, coding, testing, distributing, and using divided among people? Which materials and discourses make up software production and use? Who benefits from its production? How are ideal users constructed through its design, and how do concrete users deny that construction? This discussion takes us into the realm of software engineering.

Since the late 1960s, software production has been dominated by the engineering metaphor. Software producers first debated, and then took up, the metaphor at a 1968 NATO-sponsored conference in Garmisch, Germany, titled "Software Engineering." As the conference proceedings note, "The phrase 'software engineering' was deliberately chosen as being provocative, in implying the need for software manufacture to be based on the types of theoretical foundations and practical disciplines, that are traditional in the established branches of engineering."[28] The engineering metaphor provided software producers with methodologies from more established fields, such as metallurgical and electrical engineering. In addition, rather than conceiving of software production as an art form or the production of knowledge for knowledge's sake, the engineering metaphor oriented producers toward making objects for use and thus for users/customers. This producer/consumer orientation was compatible with capitalism and with software's newfound independence from hardware (largely because of IBM's "unbundling" of the software previously included when customers leased its machines).[29] Also compatible with capitalism was the engineering metaphor's emphasis on the management of labor; as it is described in professional documents, such as the *Guide to the Software Engineering Body of Knowledge*,[30] and by such organizations as the Institute of Electrical and Electronics Engineers (IEEE), the Association for Computing Machinery (ACM), and the Accreditation Board for Engineering and Technology (ABET), engineering post-Garmisch is marked by hierarchical control and the division of labor to produce software commodities.[31]

To be sure, there have been alternatives and challenges to the engineering metaphor.[32] However, it does provide software studies with a useful language with which to perform critical analysis. In a chapter in a key text for the field, *Software Studies: A Lexicon*, Andrew Goffey notes, "Software engineering . . . concerned as it is with the pragmatic efficacy of building software for particular purposes, might appear to offer a better starting point for factoring culture back into software."[33] Goffey goes on to note that the basics of software engineering must be explored to complete this factoring. Similarly, Nick Montfort and Ian Bogost argue that software engineering's emphasis on code and the organization of labor in the production of code provides a rich set of concepts for a critical study of software platforms.[34] Indeed, following Kitchin and Dodge, if software studies "focuses on the etiology of code,"[35] then considering a dominant paradigm of software production—software engineering—should be fruitful. Throughout this book, I use such software engineering concepts as *architecture*, *implementation*, *abstraction*, and *decomposition* as entry points into social media software, specifically in terms of the ways in which social media software is produced, the relationship between producers and users, and the ways in which users are allowed to produce the content within social media software frames.

Reverse Engineering

The next engineering metaphor I draw on is reverse engineering. Kathryn Ingle offers a brief definition of this process: "If forward engineering is the traditional process of moving from high-level concepts and abstractions to the logical, implementation-independent design needed in a physical system, then reverse engineering is the design analysis of the system components and their interrelationships within the higher-level discrete system."[36] While software engineering is the production of an abstract architecture followed by its implementation, reverse engineering starts with the final, implemented product and takes it apart, seeking clues as to why it was put together in the way it was and how it fits into an overall architecture.

I see three good reasons to engage with this metaphor. First, reverse engineering helps when we are confronted with closed code and proprietary formats. As such, it is a practical move in an age of Digital Rights Management (DRM), compiled binaries, and vicious lawsuits. If we are denied access to source code, or if software is running on a far-off server, we have to speculate on how a particular instance of software works. Drawing on knowledge of software engineering, we can do this by theorizing which sorts of uses and requirements the software is meant to meet and how the architecture and internal implementation of it might meet those needs. These may be visible in the interface; in this case, using the artifact and seeing how it shapes uses (and therefore users) through its design elements can lead us to speculate as to why some uses are privileged while other technically and equally possible uses are denied. Outside the interface, we can also turn to white papers, user manuals, press releases, blog posts, and news stories to glean insights into why coders and architects built their particular closed software system and what their goals may have been. Reverse engineers have used all these techniques for decades.

Second, to reverse engineer is also to move back through time. Good reverse engineering takes temporality—if not history—seriously. "In spanning the life-cycle stages," write Elliot Chikofsky and James Cross, "reverse engineering covers a broad range starting from the existing implementation, recapturing or recreating the design, and deciphering the requirements actually implemented in the subject system."[37] Reverse engineering is thus a forensic process,[38] taking small parts of a system and theorizing as to which abstract purposes their creators may have intended. Concrete, particular instances of software are the starting points for analysis that drives toward an understanding of larger sociotechnical systems that precede it. And, in addition to tracing the associations from concrete implementation back to abstract architecture, this emphasis on temporality also helps us deconstruct the metaphors and language of social media and technology. Many scholars and popular writers place far too much emphasis on newness and novelty. Such terms as "new media" and especially "Web 2.0" posit a radical break with the past, eliding—even denying—history, even though the techno-utopian worship of the new has been with us for at least all of modernity.[39] This sort of thinking gives rise to the breathless proclamations of "revolutions" of "new technologies." However, if we take the reverse engineering metaphor to heart, we ignore grand "revolutions" and instead do the practical work of tracing technologies back through time to uncover their associations with prior technologies and practices. The names and descriptions of current technologies often draw metaphorically on older ones, and tracing these metaphors back through time reveals hidden histories that get lost when we declare that our current time represents a radical break with the past.

Finally, reverse engineering provides a healthy perspective on the well-worn (and yet always compelling) debate about structure and agency, particularly in terms of technology's relationship to agency. If one takes reverse engineering seriously, "every time you have a better idea you will consider all the positive design aspects before condemning an entire product."[40] This approach is pragmatic, dealing with the technology we actually have rather than dreaming of utopia. Contemporary social media have many advantages over older forms of mass media: protest movements, such as the Arab Spring and Occupy Wall Street, have used Twitter, Facebook, and YouTube to organize their movements and promote their ideas when other traditional media organizations have overlooked them. We have to acknowledge these progressive facets of social media, but we do not have to accept the system wholesale. Reverse engineering is useful here, because it is a critical dissection of existing technology with the goal of building a better system. While "positive design aspects" are not essential properties of a technology—that is, they are a matter of context and social structure—there is no reason they cannot be excised from the current architecture and shaped into new forms of media more compatible with radical democracy. I find reverse engineering's emphasis on actually existing, material technology useful to this end.

Heterogeneous Engineering

So now we have a few ways to answer some of the provocations raised by sumoto.iki's work, and we have a place to start for a critical study of software centering on the software engineering metaphor. However, we have to place these post-human frames and the processes of their production within an existing political-economic human context. To do so, we need a theory and a normative stance. Here, I turn to the science and technology studies theory of heterogeneous engineering, a subset of actor-network theory[41] that is certainly useful to software studies.[42] Perhaps the biggest champion of the theory of heterogeneous engineering as well as the provider of a heuristic method to apply that theory is John Law. Law argues that social analysis should start with the "metaphor of *heterogeneous network* . . . a way of suggesting that society, organizations, agents, and machines are all *effects* generated in patterned networks of diverse (not simply human) materials."[43] Law wants to break concrete social and technological totalities into their constituent elements to discover "the complexity and contingency of the ways in which these elements interrelate" and document "the way in which solutions are forged in situations of conflict."[44] "Solutions" in heterogeneous engineering are not simply the best possible forms of technology or techniques for the job at hand. Rather, they are "associations" of technological and cultural elements, joined together despite myriad forces that constantly threaten to *dissociate* the object in question. Associated elements are "difficult to tame or difficult to hold in place. Vigilance and surveillance have to be maintained, or else the elements will fall out of line and the network will start to crumble."[45] Every heterogeneous element is resistant to being put to the intended purpose, and anyone claiming to be an engineer must recognize the agency of the nonhuman.

Thus, the task of the critic is to start with the totality, seek out its constituent processes, and articulate the connections and relationships among them to (re)build a complex-concrete whole. The critic must seek out processes that are hidden, implied, or on the margins and connect them to those that are obvious and privileged in commonsense discussions of the object. Heterogeneous engineering is especially valuable for our purposes here in its conceptual grasp of technology; it reminds us (1) of the highly contingent nature of any technological artifact; (2) that "knowledge" and "technique" are the result "of a lot of hard work in which heterogeneous bits and pieces—[such as] test tubes, reagents, organisms, skilled hands, scanning electron microscopes, radiation monitors, other scientists, articles, computer terminals, and all the rest—that would like to make off on their own are juxtaposed into a patterned network which overcomes their resistance";[46] and (3) that very often these heterogeneous bits and pieces must be *engineered* by some actor or organization.[47] The heterogeneous engineer must be an engineer of *différance*, not only associating the obvious elements but also deferring, eliding, displacing, and denying deviant elements, the Other, heteroclites, and noncoherences. The heterogeneous engineer works with the immaterial and material, the abstract and the concrete, none of which is easily or necessarily associated together. If the engineer is good enough, then the system appears to be second nature, given, and immutable; the contradictions and resistances within the materials it comprises are hidden, and we celebrate the system as evidence of linear technological progress. Very often history is told as the story of these successful projects, while failures and contradictions (that is, messy phenomena that trouble our lovely narrative of linear and logical technological progress) are pushed outside the margins of historiography.

This theory reminds us that we cannot simply uncritically accept the proclamations or products of software (or reverse) engineers. We have to empirically examine them, produce litanies[48] of human and nonhuman objects, and critically trace their associations. We have to ask what is hidden within these engineering metaphors: beneath the surface of the interfaces driven by Asynchronous Javascript and XML (AJAX),[49] among the racks of server farms, inside client devices, in the wires and radio signals of network connections, and between the lines of IPO-filing documents, roadshow proclamations, and tech fan blogs. To analyze social media software and culture, we have to look forward and backward, in the abstract and concrete, in the now and in the vestiges of the past. We see contradictions between reverse and "forward" software engineering: there are competing movements through time, associations and dissociations, dreams of future solutions and the frustrating structures of the past. We find moments when the software misbehaves, when a security hole appears and a viral infection sneaks in, or when the subject so often Othered by engineering—the user—asserts him- or herself and disrupts the system. Heterogeneous engineering troubles the neat claims to agency made by engineers by hinting at the agency of nonhuman elements, such as machines and code. Heterogeneous engineering reminds us that an electron out of place can doom a network or that a shift in social politics can cause a reinterpretation of the ontological reality of a machine. Ultimately, then, this theory acts as a counterweight to the triumphalism of the software engineering literature.

A Normative Intermezzo

Finally, before I turn to chapter summaries, I need to take a moment to lay out a normative position. Because hegemonic social media is produced within informational capitalism, the inherent inequalities of that system are translated, ported, or simply replicated within new media software. As a growing number of media scholars, activists, and social media users now recognize, new media capitalism as practiced by such sites as Facebook, Google, and Twitter has had terrible consequences: it reduces online interaction to binary declarations of like-it-or-not consumer choices; it hides and inures us to the surveillance systems operating underneath its surface; it relies on the free labor of its users to build its content while the site owners make billions by selling user data and stock in their companies; its centralization (which always seems to be part of the political economy of capitalism, despite the repeated cries of creative destruction, disruption, innovation, and competition) provides an all-too-easy means by which states can gather data on citizens; and it promotes a culture of anxiety and immediacy over depth. In short, contemporary social media hardly seems compatible with democracy, and because of this issue, I need to move past the point where heterogeneous engineering analyses tend to leave off:[50] at the question of "what is to be done?" As Marx has famously argued, "The philosophers have only interpreted the world, in various ways; the point is to change it."[51] After we have explored software engineering, reversed it back through histories of computer science and modernity, and discovered the hidden Others and forces that threaten to dissociate that totality, we then may glimpse new potential associations of processes, technologies, and subjectivities. We begin to realize that these new associations can be heterogeneously engineered to reflect different values—specifically, those of the much-longed-for Habermasian public sphere of democratic debate and decision making. We start to see paths forward out of the seemingly unassailable realms of power into new spaces of freedom. Thus I turn, at last, to the normative stance of critical political economy of communication, particularly the Marxian strain, as I analyze the engineering of social media software. I also rely on critical Marxian analysis as I reverse engineer this system and look for something better.

Thus, this book critiques social media by rigorously analyzing the ideas of social media engineers, reversing them to trace the path from concrete software to abstract desires, seeking contradictions among them, noting where meanings shift as audiences and interests shift, contrasting them with the underlying architecture, and looking for new, progressive possibilities and shapes within this structure.

The Plan of the Book

Chapter 1 begins this process by exploring the emerging phenomenon of the socialbot. Throughout many critical examinations of social media, there is a nagging sense that human activity is reduced to datasets within the templates of such sites as Facebook, Twitter, and Google. Socialbots—automated social media profiles that tweet, like, and friend—are concrete evidence that this nagging feeling is not unfounded. Socialbots are a reflection of our activities within social media; for these machines to work, we ourselves have to be trained to be machinelike as we produce images of ourselves via publicly confessed declarations. In keeping with the idea of reverse engineering as looking back, this chapter links contemporary socialbots to the construction of the ur-socialbot, Alan Turing's Universal Machine, a seminal idea in early computing. Turing's machine and his famous Turing Test of computer intelligence are founded on his idea that the human mind can be understood as a finite, discrete machine. As such, if various "states of mind" of humans can be encoded, they can be manipulated and imitated by computers, even to the point where a computer could appear intelligent by engaging in a conversation with a human. Chapter 1 then links Turing's work with the emerging critical literature on "noopolitics" (the politics of *nous*, or the mind). Next, it explores the cultures of transparency and authenticity in contemporary social media. Social media's "culture of confession" is producing a massive dataset of the internal, discrete states of mind of human beings. Once this codification is done, then the Universal Machine (in this case, socialbots) can imitate the human in a modern-day Turing

Test. Socialbots' success in this regard certainly raises the specter of machine intelligence. But more importantly, socialbots are emerging as a technology of noopower (i.e., power over thoughts) capable of shaping opinions and interactions within social media.

In Chapter 2, I explore a power imbalance between social media users and owners, particularly as the two meet in the "Web as platform." Looking backward once again, I link a division of labor apparent in social media to the internal divisions of labor built into computers, best exemplified by the Von Neumann Architecture. This architecture, which divides the processor from memory and allows the storage of programs, has been replicated in social media. In social media, users are granted the power of the processor: they are allowed to process digital objects, liking this, tweeting that, rating, commenting, and sharing. This is "collective intelligence" in action—at least as that intelligence is modulated within social media. However, social media companies maintain a strict division between the user/processor and memory, the other side of the computational equation. Social media companies derive power from storing the results of users' affective processing in archives. These archives can then be drawn on to construct "facts" about users: which consumer objects they desire, which trends they are tuned into, even whether they might agitate for political change. These "facts"—artifacts of a particular "harnessing" of collective intelligence—can then be sold to marketers or surrendered to states. This division between the processor and the archive helps maintain the social structure of social media.

Chapter 3 continues on the theme of the division of labor, this time considering how the software engineering practice of abstraction has shaped the processes of user labor. The central questions I consider here: how "user-led" is user-led production? Is there a limit to what users can do within social media? Who sets that limit? To answer these questions, I propose the concept of the "real software abstraction," a synthesis of the software engineering practice of abstraction and Marx's concept of the real abstraction. Starting with the seminal work of software engineer Frederick Brooks, I argue that the predominant architecture of social media has been drawn from the methods of managing labor in the production of software. Brooks's work on the IBM System/360, detailed in his book *The Mythical Man-Month*, was based on his division between architecture and implementation. Architecture is a high-level abstraction, the blueprint of the system. Implementation is the labor of coding to the architectural specs. Drawing on the work of Alfred Sohn-Rethel and Alberto Toscano, I link architecture to Marx's real abstraction, arguing that software architecture may be an abstraction, but it has real, concrete effects in the world, particularly in how it enables the disciplining of coding labor. I use a case study of Myspace and Facebook to illustrate instances where this practice has been transferred from firm-based software production to social media. Social media site owners create real software abstractions and allow their implementation to be carried out by the free labor of users.

Chapter 4 is part of a larger conversation about networking protocols and standards. Many narratives discussing the rise of the Internet and Web as new forms of media center on the production of shared standards and protocols, such as Transmission Control Protocol/Internet Protocol (TCP/IP) and Hypertext Transfer Protocol (HTTP). In turn, social media are seen to be animated by the Application Programming Interfaces (APIs) of Google, Facebook, and Twitter. In Chapter 4, I complicate this narrative of standards by looking at the history of *advertising* standards, especially those developed by the Interactive Advertising Bureau (IAB). Our current social media have been designed to adhere to the standards produced by the IAB. Through these standards, social media sites link heterogeneous user-generated content of all kinds (from posts to status updates to the affective labor of constituting social networks) to networks of marketing and advertising. Advertising standards have thus played a major role in shaping the surveillance-based business models of such social media sites as Facebook, Twitter, and Google. In addition, in Chapter 4, I synthesize many of the arguments made in the previous chapters by outlining the subject imagined by the IAB and marketing-centric social media: the *sovereign interactive consumer*. This subject is the prime abstraction of social media. The IAB and social media sites build their discourses of self-regulation on this abstract consumer.

Of course, these chapters are quite pessimistic. As a way out, there is the normative intervention to consider: what is to be done? The remaining chapters use the contradictions and gaps exposed in the previous ones to plot actual and potential ways forward. Chapter 5 details Wikipedia's evolution. This now-famous encyclopedia relies on many of the processes underpinning social media as a whole: it harnesses collective intelligence, it provides a platform on which users can build new applications, and it relies on a client-server architecture. And yet, Wikipedia is radically different from other social media sites. This chapter explores Wikipedia's difference in terms of an important event early in its history, the "Spanish Fork." In 2002, Wikipedia users in Spain left the site in droves in response to the potential use of advertising on the site. The leaders of the Spanish Fork believed that advertising would reflect Wikipedia's exploitation of their labor, so they transplanted the entire Spanish-language version of Wikipedia to their own servers hosted at the University of Seville. Clearly, this Spanish Fork was a labor strike; the participants in the strike were able to heterogeneously engineer a "class for themselves" out of the typical social media capitalist production of the user "class in itself." Their encyclopedia, *La Enciclopedia Libre Universal en Español*, became a rival to the nascent Wikipedia, forcing Wikipedia's founders to establish it as a nonprofit site rather than as the commercial venture they originally intended. I see this historical moment as proof that, with the right mix of technological and communication infrastructure, leadership, and discourse, users can recognize their position as free laborers and strike against social media exploitation.

Finally, in Chapter 6, I return to the reverse engineering metaphor. In the literature on reverse engineering, the impulse to do the work of disassembly and documentation is to improve the bottom line: products can be improved, technology and infrastructure can be maintained instead of thrown out when they break down, and new intellectual property can be claimed. In contrast, as a critical student of online media, I aim my work toward improving our media system, which is a fundamental element to democracy and citizenship. To this end, I offer a "Manifesto for Socialized Media" in two parts. The first is a design specification—or, to be more truthful, a *design speculation*—for a socialized media system that is decentralized, transparent, encrypted, antiarchival, stored on free hardware, and geared toward collective politics over atomization and depth over immediacy and surfaces. The second part is a brief survey of current efforts to implement and realize such a system by heterogeneously engineering social media alternatives. Because I play the role of a designer, my ideal socialized media system remains just that—an ideal. However, looking at the production of media systems meant for activists, it is clear that serious efforts are underway to realize an ideal system, such as the one I have proposed. Moreover, because software is obdurate, and because it often asserts its agency by shaping and determining actions, the encoding of progressive politics into socialized media systems is also the preservation and extension of antistate, antipower, and anticapital politics.

This is the plan of the work. However, even after all this, social media will no doubt appear to be a complex, almost overwhelming phenomenon—and moreover, by the time this book appears, no doubt many elements will have changed! As Geert Lovink admits, "The object of study is in a permanent state of flux and will disappear shortly—the death of everything cannot be denied."[52] Book-based case studies and analyses of networked media, however timely, fall behind network time. No book or single author can hope to grasp social media in its entirety. But Lovink suggests a "possible way out" of this quandary: "the development of critical concepts that migrate from one generation of applications to the next, without falling back on speculative theory that merely celebrates the liberating potentialities of buzzwords while waiting to be translated into market value."[53] I hope the process of reverse (heterogeneous) engineering and its associated concepts I offer here are a contribution to the work of activists, academics, and users who seek to change social media for the better.

1

The Computerized Socialbot Turing Test

Noopower and the Social Media State(s) of Mind

> Our machines are disturbingly lively, and we ourselves
> are frighteningly inert.
>
> —DONNA HARAWAY, "A Manifesto for Cyborgs: Science,
> Technology, and Socialist Feminism in the 1980s"

The last tweet you got may have been from a robot.

Networks of socialbots are beginning to spread across social media. Internet users have long been familiar with bots;[1] the most benign ones are Web crawlers that index sites for search engines. Wikipedia editors may have seen some of their edits cleaned up by editing bots. Automated searches and e-mail sorting are another type of bot built into e-mail clients. However, socialbots are different from their more benign predecessors. According to Tim Hwang, Ian Pearce, and Max Nanis, "What distinguishes these 'social' bots from their historical predecessors is a focus on creating substantive relationships among human users . . . and shaping the aggregate social behavior and patterns of relationships between groups of users online."[2] That is, as their name implies, socialbots are built to be *social*, to interact with us while we are using Facebook and Twitter. They also work to subtly alter how social media users interact with and link to one another.

Moreover, to more effectively shape online behavior, according to Y. Boshmaf et al., "What makes a socialbot different from self-declared bots (e.g., Twitter bots that post up-to-date weather forecasts) and spambots is that *it is designed to be stealthy; that is, it is able to pass itself off as a human being."[3]* They appear to be fellow humans, with profiles, avatars, and status updates, who respond to direct messages and questions from other social media users. Although software engineering research into socialbot construction is in its infancy, some have already been successful, such as James M. Titus, a Twitter socialbot that won the 2011 Socialbot contest by gaining more followers and direct messages than other competitors.[4] Similarly, a research team from the University of British Columbia built a socialbot network that was able to befriend humans on Facebook and gathered more than 250 gigabytes of private user data.[5] Finally, the U.S. Air Force is using "persona management," a tactic of managing multiple fake accounts on social networks to infiltrate terrorist cells.[6]

An immediate reaction one might have to the possibility that socialbots are manipulating us or gathering information on us via Twitter and Facebook is unease, especially since it is hard to tell a socialbot from a human being. It is unnerving to think that the last heartfelt tweet posted by a celebrity on Twitter might just be algorithmically selected from a preapproved list or that our latest friend request ultimately came from the Department of Defense. However, in this work, I am less interested in the ethics of socialbots; instead, here I explore how socialbots expose a part of the larger project of software engineering of social media and ultimately how these bots reveal a part of social media's (noo)political economy of surveillance, attention, and modulation. Throughout many critical examinations of social media, there is a nagging sense that human activity is reduced to datasets within the templates of such sites as Facebook, Twitter, and Google. Socialbots are concrete evidence that this nagging feeling is not unfounded—that, in fact, a radical reduction of human activities is happening in social media, and, moreover, this reduction serves to maintain and extend the (noo)power of social media site owners, third-party marketing firms, and government surveillance systems. As Karl Marx argues, "The *devaluation* of the world of men is in direct proportion to the *increasing value* of the world of things."[7] In other words, the more we produce, the less we are, and although it is counterintuitive, this idea still holds in a media system where we are encouraged to produce *ourselves* via declarations of our desires and connections. Socialbots are a reflection of our activities within social media; for these machines to work, we ourselves have to be trained to be machinelike. In sum, the construction of socialbots is evidence of social media software engineering as social engineering in the age of noopolitics.

To examine this, I first explore the construction of the ur-socialbot, Alan Turing's Universal Machine. I argue that Turing's machine and his famous Turing Test of computer intelligence are founded on his idea that the human mind can be understood as a finite, discrete state machine. As such, if the various "states of mind" of humans can be encoded, they can be manipulated and imitated by computers, even to the point where a computer could appear intelligent by engaging in a conversation with a human.

I then connect this history to the concept of noopolitics. "Noopolitics" is a term coined by Maurizio Lazzarato to describe our contemporary emphasis on the politics of attention and memory.[8] "Noo" derives from *nous*, the Greek word for mind. In Lazzarato's view, the older political forms that Michel Foucault describes (discipline and biopolitics) are being subsumed into the politics of attention and memory.

Next, I briefly explore transparency and authenticity in social media. I suggest that social media's "culture of confession" is producing a massive dataset of the internal states of mind in human beings. Once this codification is done, then the Universal Machine (in this case, socialbots) can imitate the human in a modern-day Turing Test. Socialbots' success in this regard certainly raises the specter of machine intelligence. But, more importantly, socialbots' success demonstrates that social media capitalism is getting more and more skilled at alienating the fruits of laborers—in this case, data appropriated from social media users via surveillance and standardization. Ultimately, I argue that this alienation serves a particular end. Social media is an instantiation—albeit a nascent one—of noopower: the action before action that works to shape, modulate, and attenuate the attention and memory of subjects.

Universal Machine Transparency: Turing's Famous Machine and His More Famous Test

Socialbots are the latest in a long line of clever software meant to fool humans. To reverse engineer them, we have look back to the work of Turing and his Universal Machine. The Universal Machine first appears in Turing's famous 1936 paper, "On Computable Numbers, with an Application to the *Entscheidungsproblem*." This paper is seminal for two reasons: it introduces an abstract, digital computer, and it demonstrates the theoretical, mathematical limits of what this machine could effectively do. The second point, which is based in part on the *Entscheidungsproblem* (the "decision problem"), is rooted in the mathematical theories of David Hilbert, Alfred North Whitehead and Bertrand Russell, and Kurt Gödel and is not my concern here.[9] Suffice it to say, on the basis of Gödel's incompleteness theorem, Turing is able to define in the abstract a machine that could compute numbers and solve problems within a formal system (while bracketing off those problems that could not be proven within the formal system). This machine is, of course, the Universal Machine, which is my focus here.

At the heart of the universal machine is a binary system. In Charles Siefe's description, the machine

> reads information from a[n] [infinitely long] tape. This tape is divided into squares that are either blank or have a mark written on them. A Turing machine is extremely simple. It can only perform a handful of basic functions: read what's on the tape at a given position, advance the tape or rewind it, and write or erase a mark on the tape.[10]

Turing's machine thus has a small set of logical operations (e.g., read, write, erase, advance, and rewind). However, despite the simplicity of this machine, it can compute any sequence of complicated algorithms or formulas by building more complex operations out of those simple building blocks. The only requirements are that the instructions be encoded into a form the machine can read and that the algorithms involved fit within the theoretical limits on mathematical systems outlined by Gödel. Thus, in theory, any other machine's operation can be controlled or replicated by the Universal Machine, so long as each state of the target machine is discrete and is written in a standard format. Although "strictly speaking," as Turing writes, "there are no such machines . . . [e]verything really moves continuously,"[11] because of our modern quantification of fluid phenomena, such as time and space, we can divide continuous actions into discrete moments and thus decompose any continuous machine into a finite collection of states. And, since any state has as its binary opposite a nonstate, the binary universal machine can be programmed to replicate any discrete state machine.

Thus, any other machine or process whose states can be divided into distinct stages is transparent to the Universal Machine. One key example is human computation.[12] During Turing's time, the machinelike aspects of human computation could be found in the rooms of mainly women who operated in a division of labor to complete small parts of large equations, processes that had been organized and standardized in books and by factory-like management.[13] Turing imagines that these human operations could be replaced by a machine: as Turing notes, "If one wanted to make a machine mimic the behavior of the human computer in some complex operation one has to ask him[14] [sic] how it is done, and then translate the answer into the form of an instruction table."[15] These tables are, of course, programs. We can easily imagine human computation being translated into digital computer programs, because this is exactly what has happened.[16]

However, and most importantly, Turing does not stop there: he expands the vision of the Universal Machine from standardized human computation or simple discrete state machines to include the decidedly unmachinelike processes of *intelligence and thought* by suggesting that any machine that can imitate a human during a conversation must be considered intelligent. This is the famous "Turing Test," during which an interrogator questions two entities and tries to decide which is human and which is the computer on the basis of a short conversation.[17] As Turing explains the test in a 1952 BBC interview:

The idea of the test is that the machine has to try and pretend to be a man [*sic*], by answering questions put to it, and it will only pass if the pretense is reasonably convincing. A considerable proportion of a jury, who should not be expert about machines, must be taken in by the pretense. They aren't allowed to see the machine itself—that would make it too easy. So the machine is kept in a far away room and the jury are allowed to ask it questions, which are transmitted through to it: it sends back a typewritten answer.[18]

By the year 2000, Turing predicts in his work, computers would be powerful enough to fool 70 percent of the judges.[19] The accuracy of this prediction is not my concern here, but I do want to emphasize that Turing's game is a game of percentages and probabilities, a point I return to later.

A conversation, in this view, is a discrete process, with two entities (an interrogator and a machine or an interrogator and a human) exchanging finite strings of encoded text. Drawing on information theory, we can say that blocks of text comprising conversations are patterns purposely selected to convey messages. As such, Turing's argument that a computer with a large-enough memory and a clever set of instructions could converse with a human and fool the human into believing he or she is talking to another person can be viewed as an informational problem.[20] Moreover, the conversation is extremely limited, with no visual or auditory cues; only text is allowed.

Turing's Test, proposed in 1950, has since spawned many attempts to build intelligent chat bots. A famous early example is Joseph Weizenbaum's ELIZA, an emulation of a Rogerian psychoanalyst, which was successful in carrying on text-based conversations by using the tactic of responding to any user input with a question.[21] In many ways, Kenneth Colby's PARRY was a response to ELIZA; PARRY emulated a paranoiac, using a psychological disorder as a means for human interrogators to explain away non sequiturs in any conversation with that bot.[22] The adoption of the Internet led to many bots living in virtual worlds, such as the text-based TinyMUD, some of which have been adapted to take on the Turing Test.[23] In 1991, inventor Hugh Loebner sponsored the Loebner Prize, which has been held annually to test programmers' skills in constructing programs that can pass the Turing Test. On the basis of Turing's writings, to win the contest, a bot must be indistinguishable from a human in conversation, a feat that has not happened yet.

It is important to note that, for Turing and his interlocutors, the human mind is conceived of as a digital computer.[24] Consider the phrase "states of mind" as an entry point into this idea,[25] where because "any step of the process being followed must be describable in terms of observation of symbols, changing of symbols, and change of state of mind,"[26] the mind's functions can be textually encoded and replicated by the Universal Machine.[27] So conceived, the mind then becomes transparent, its inner workings made visible to computation—at least as evidenced by the textual descriptions of steps and states given by people carrying out myriad tasks.[28] If, as Sherry Turkle argues, computers are our new "objects to think with,"[29] then what we seem to be thinking about with them is how we go about thinking. If Turing is right, our very minds can be conceived of as discrete state machines potentially imitated by computers.

Turing's suggestion that the processes of thought and intelligence can be codified and thus made transparent to the Universal Machine is by no means the limit of the desire to quantify seemingly unquantifiable systems. His Universal Machine theory has had a nearly immeasurable impact on theoretical computer science, linguistic philosophy,[30] cognitive science, biology,[31] cybernetics, genetics, and systems theories. By conceiving of their objects as consisting of discrete states, these theories seek to bring their objects into the purview of the Universal Machine; many of Turing's interlocutors in these fields dream of truly *universalizing* philosophies.[32]

Thus, Turing's Universal Machine has led to, or at least anticipated, many modern dreams: that our language processes can be mapped and fully known, that our economy's inner workings can become completely transparent and thus made more efficient, that our weather systems can be predicted, that human behavior can be uncovered and modified with social scientific inquiry, that our neurological systems can be networked and programmed. Turing's dream—at least as it has been modified and extended in modernity—is similar to the dreams of Pierre-Simon Laplace or Frederick Taylor: that human activity, from its physical to its mental manifestations, can be known, measured, abstracted, and controlled. Indeed, examples of this dream abound and recur: consider Kevin Kelly, the executive editor of *Wired Magazine*, who argues in a best-selling business manual that we, society, and the universe are all computers, and thus whoever controls the software controls the whole show.[33]

To be fair, Turing himself probably would laugh at the vast claims to universal knowledge made in such fields as (the new) economics; after all, he is critical of the Laplacian philosophy of a predictable, mechanistic universe,[34] and one fundamental aspect of his seminal essay "On Computable Numbers" is precisely that such systems can never be complete (and thus perfect prediction is never possible). However, reading Turing closely, we see the remarkable idea that computers can imitate any other discrete state machine; this is the heart of their incredible flexibility. Moreover—and this is the point I want to emphasize—this is the heart of the concept of radical machine transparency: if the workings of a machine (biological or otherwise) can be known (measured, quantified, and notated in a standardized manner), then that target machine is transparent to Turing's Universal Machine.

Turing and Noopolitics

In addition, clearly the work of Turing, the transition from human to electronic computation, and above all the desire to encode discrete "states of mind" should be read as early noopolitical phenomena. Lazzarato and subsequent theorists contrast this form of politics with those Foucault describes as disciplinary and biopolitical. In discipline, the object is the individual body. The body is trained to work with particular instruments (e.g., pens, guns, machines). This training features the increasing, painstaking granularity of motion (moving the pen while sitting upright precisely eight inches from a desk; holding the gun while lifting the left leg to a certain height; working the drill press so the bit drills here, here, and here) and the increasing granularity of time (it should take no more than forty-five seconds to write this sentence; after the gun crosses the center of the body, the right heel comes down; thirty-five holes should be produced in a minute). Discipline is also based on enclosures: schools, prisons, barracks, and hospitals form hermetically sealed spaces in which bodies are trained. The outside is cordoned off: one's family is of no concern to the prison warden, while thoughts of factory work are not brought with the body into the school. Within these enclosed spaces, pupils, prisoners, students, soldiers, and workers are arranged in hierarchies both physically in terms of location within space and ordinally in record books.

Foucault's biopolitics extends this logic to the enclosed spaces of the nation-state. Rather than working on the individual body, biopolitics takes as its object populations and works on this object via metrics of life and death (e.g., birth rates, disease statistics, productivity metrics). The enclosed body politic is secured against threats to its health, such as Others beyond the borders, from viruses to foreign armies and populations, by securitization practices. Within this enclosure, liberal policies, such as welfare, social security, and national healthcare, help maintain a productive body politic.

In his development of noopolitics, Lazzarato concurs with Gilles Deleuze's argument that Foucault's description of enclosure, discipline, and biopolitics is of a historical moment (largely describing Western states in the nineteenth and twentieth centuries) rather than of contemporary society (broadly speaking, after World War II). For Deleuze, the subjects imagined by discipline and biopower have escaped enclosure: "Everyone knows that [such institutions as the family, prison, hospital, school, and factory] are finished, whatever the length of their expiration periods. It's only a matter of administering their last rites and of keeping people employed until the installation of the new forces knocking at the door."[35] These "new forces" are the modulated flows of control that seek to exploit the virtuality and possibilities of becoming that exist outside enclosure. The objects of control are exchange rates, publics, global flows of commodities, and, above all, *minds*. As Lazzarato argues, "If disciplines molded bodies by constituting habits mainly in bodily memory, the societies of control modulate brains and constitute habits mainly in spiritual memory."[36]

Drawing on Gabriel Tarde, Lazzarato sees memory as constitutive of life itself:

> According to Tarde, without memory, without this force (a duration that conserves), without this fertile succession that contracts the before in the after, there would be no sensation, no life, no time, no accumulation and thus no growth. For Bergson, Tarde's first "disciple," without this duration the world would be forced to start anew at every moment. The world would be a present repeating itself indefinitely, always equal to itself. Matter itself would not be possible without this duration. The creation and realisation of the sensible presuppose the activity of memory and attention, as well as their power of actualisation and repetition.[37]

As such, memory's constitutive power has been, in Lazzarato's view, the *outside* of the enclosed spaces of discipline and biopower, the virtuality and becoming that enclosure seeks to cordon off: "Memory, attention and the relations whereby they are actualised become social and economic forces that must be captured in order to control and exploit the assemblage of difference and repetition. It is by remaining faithful to this intellectual tradition that Deleuze can affirm that in 'a life there are nothing but virtuals.'"[38]

And here, then, we see the possibilities of noopower, the institutionalization of the politics of mind, wherein (to use Foucault's definition of power)[39] one mind's action shapes other minds' actions: memory and attention. Noopower, then, is a relation of power in which one mind may act before others, and, as we shall see, this is often done at a distance. Such acting-before-acting minds aggregate and codify the possibilities of their actions by forming institutions: mass media, polling companies, market research firms, and states interested in shaping public opinion. Noopower is institutionalized thought that (to borrow from Foucault) "incites. . . . [I]t induces, it seduces, it makes easier or more difficult; it releases or contrives, makes more probable or less; in the extreme, it constrains or forbids absolutely, but it is always a way of acting on one or more acting subjects by virtue of their acting or being capable of action."[40] Such power is a way to modulate and condition the sheer potentiality of immaterial (i.e., cognitive) labor power, to subtly shift the probabilities that the public will think in particular ways, support particular policies, or consume goods in particular patterns.[41]

As such, Turing's interest in encoding and programming "discrete states of mind" into machines is an early move in the computerization of noopolitics. By considering how intelligent minds and digital computers could be linked, Turing is a noopolitical pioneer who sees the possibility that even thoughts could be programmed. Theorists of noopolitics, however, have largely ignored Turing's work. This may be because the Turing Test largely pits the intelligence of an individual machine against a human judge, and noopolitics and noopower involve *masses*—or more precisely, using Tarde's distinction, *publics*. The question is this: could Turing's intelligent machine become a noopolitical agent, working on publics? The recent rise of socialbots provides an answer: yes.

The Social Media Confessional Machine

Transparency is all the rage these days.

There are countless calls for transparency—that is, radical exposure of internal details—for a wide range of entities. Government agencies must file more progress reports. In turn, states demand that corporations produce reports on their finances and activities. Hospitals and healthcare providers are subject to online consumer ratings sites. Professors in higher education must quantify the learning their students have done, and universities are being pressured to share these assessments with the public.

This push for transparency extends to, and in many ways is animated by, digitally mediated communication. In addition to large organizations exposing their finances, internal practices, and systems of expertise production, individuals are turning transparent. In media studies, much has been said about the "culture of confession" we live in.[42] Confession—the revelation of private personal details to another party—is now often linked to tabloidization of the news,[43] talk shows, and reality television.[44] As such, as Nick Couldry argues, confession has become a "media ritual," a practice that reinforces the power of media in our daily lives. In Couldry's view, the media are powerful because they provide a potential space for "ordinary people" to enter and become "extraordinary."[45] Ordinary people can do so by getting in front of cameras and "getting real": presenting their lives for all to see and revealing their internal emotional states via performances.[46] Ultimately, by confessing private details, participants in reality television or talk shows deny the mystification of mass mediation by exposing such elements as professional editing and the aura of the camera as socially constructed.[47] This represents a somewhat perverse modification of the feminist slogan: the personal is no longer just political but must necessarily be public, exposed to allow us to function in the affective, knowledge economy.[48] Feeling and emotion are to be performed for ever-larger audiences.[49]

We might then argue that social media have amplified and extended this practice by breaking it out of the confines of mass media, which, for all the "reality" of "reality TV," are always highly constrained.[50] Indeed, social media seem to allow us to *better* become the media by allowing users to confess and reveal their emotional and subjective states without the mediating influence of mass media gatekeepers. Alice Marwick and danah boyd's work on Twitter reveals the contours of authenticity and sincerity in social media.[51] They argue that authentic tweeting is a cultural construct; to be authentic in many online cultures is to be passionate, emotional, dangerous, and intimate.[52] This is contrasted with safe, preapproved genres of discourse, such as press releases, carefully crafted advertisements, edited interviews, or lawyer-vetted public statements. Marwick and boyd argue that the more effective form of tweeting (at least as judged by the number of followers and the amount of back-and-forth discussion between the follower and the followed) is the spur of the moment, uncensored personal revelation. Whether the Twitter user is a celebrity, an "ordinary person" (to use Couldry's term), the personally branded,[53] or a microcelebrity,[54] the path to effective communication in social media appears to confession and revelation of personal details.

Perhaps the best example of the confessional is on YouTube, where the aesthetics of amateur production are dominant: YouTube vlogs often feature a single person, sitting at a computer (often located in a bedroom), gazing directly into a Web cam, and—at least in the "real vlog" genre—confessing personal details.[55] The eye contact (mediated via the Web cam and computer screen) and personal details are meant to communicate the idea that this is as unmediated as YouTube can get.

The emphasis on the cultural constructs of "authenticity" and "confession" is in part a reaction to the sheer noise of social media; for me to connect to a friend on Facebook, I must cut through the clamor of my friend's social stream. A strategy for doing so is to be "real," to express details of my personal life. If my audience is larger than a single friend, then authenticity is far more necessary to "stand out from the crowd," as the personal branding literature often puts it. "Transparency is so chic," notes an informant in one of Marwick and boyd's Twitter studies,[56] and indeed, in response to the noise of daily life, transparent confession seems refreshing.

I should note that I do not want to overemphasize this concept of the "culture of confession." In my own use of social media, I often think about what I am tweeting or posting, and I do take care to consider my own privacy and how I present myself to others. I am sure many people do the same. Indeed, boyd's research often reveals the limits of transparency and the violations of privacy that social media users often resist.[57]

However, and importantly, I also do not want to reduce the idea of confession, authenticity, or privacy to an *individual* concern. It certainly appears to be, as individual users work to personally brand themselves and become highly visible nodes in the social graph. But in the words of Pappy O'Daniel in *O Brother, Where Art Thou?*, "We ain't one-at-a-timin' here. We're mass communicating!" In other words, this is decidedly not one-by-one appropriation. What is happening is the appropriation of *masses* of data—patterns and shapes, what Tiziana Terranova (following Jean Baudrillard) calls "social entropy" of a mass that is capable of absorbing and dispersing meaning into an array of probabilities.[58]

Importantly, a major outcome of transparent, confessional social media communication is the public production of *patterns of textually encoded, discrete states of mind.* In the language of *Web 2.0 Architectures,* a Web 2.0 software engineering guide, this is "declarative living":

> In conversations around the world, people are constantly expressing their preferences and opinions. In other words, we live *declaratively.* We declare who our friends are and who are acquaintances or business colleagues are. We talk about the videos, music, books, art, food and so on that we've encountered, the people that inspire us, and the people we'd rather avoid. It's no different on the Web, except that on the Web we can make these declarations explicitly through common formats and technologies . . . , or we can leave a trail of digital breadcrumbs and let other people draw their own conclusions.[59]

Research into the "Semantic Web" centers on mapping and analyzing such declarations and "digital breadcrumbs." Given the structure of social networks, where we self-affirm, where we "like," "friend," or otherwise connect with such objects as other people, movements, ideas, brands, bands, and books, complex maps of aggregated user tastes can be produced.[60] In sum, the "big data" produced by a mass culture of confession are a boon to computational analysis.

And this is where we see the first clue of how socialbots are able to do their work: a major outcome of transparent, confessional social media communication is the public production of aggregated *patterns of textually encoded, discrete states of mind.* This is exactly the type of information that a bot needs.

It is a short path from massive, analyzed databases of digitized states of mind to machines capable of playing convincing imitation games. In the words of socialbot engineers Hwang, Pearce, and Nanis, "Digitization drives botification; the use of technology in a realm of human activity enables the creation of software to act in lieu of humans."[61] In much the same way that automation on the assembly line required the encoding of the averaged motions of industrial workers into machines, socialbots require data on the *aggre-*

gated actions of users in social media. Socialbots are thus reflections of the *typification* of masses of users; as the Twitter socialbot project Realboy puts it, "We are not impersonating Twitter users[;] we are simply imitating them."[62] Just as advertising networks, such as Yahoo! and Google, track users' movements around the Web and place them in a taxonomy of consumer types (e.g., "Sports Enthusiast," "Single Mother"), socialbots replicate patterns of user activity.

To be certain, the digitized patterns that socialbot engineers have exploited in Facebook and Twitter are quite crude. For example, socialbot engineers Boshmaf et al. note that the goal is to create socialbot profiles that are "socially attractive":

> We consider a purely adversarial standpoint concerning social attractiveness; the adversary aims to exploit certain social attributes that have shown to be effective in getting users' attention. Such attributes can be inferred from recent social engineering attacks. Specifically, using a profile picture of a good looking woman or man has had the greatest impact. Thus, an adversary can use publicly available personal pictures for the newly created profiles, with the corresponding gender and age-range. In fact, the adversary can use already-rated personal pictures from websites like hotornot.com, where users publicly post their personal pictures for others to rate their "hotness."[63]

The crude quantification and digitization of the seemingly unquantifiable social trait attractiveness on hotornot.com (a site famous for posting images of women and allowing users to rate their "hotness" on scales of 1 to 10) produces one key pattern needed for socialbots. These patterns are being built, right now, via such aggregated actions of users. Because socialbot engineering is in its infancy, these simple patterns (people with many friends are less likely to screen new requests, clever status updates appear human, the triadic closure principle holds in social media, attractive people get more friends, and so on) are no doubt the tip of the iceberg—more sophisticated patterns will soon be encoded and included in these robots.

When we are logged into Facebook or Twitter, we are not aware that we might be participating in the Turing Test; instead, we simply interact with digitized versions of each other—and potentially with bots. This points to the particularly *mediated* aspects of social media: the interactions are framed in particular ways, just as the original Turing Test is a highly delimited form of interaction. The Turing Test calls for particular limitations: the machine is to be in another room, separated from human judges, and is to interact via text. The human judges cannot see or hear the machine (or the human who is also playing the imitation game to provide a comparative conversation by which to judge the machine), nor can they speak directly to the machine. On the basis of this setup, the goal is to fool 70 percent or more of the judges. Social media operate with similar constraints: messages are sent via particular channels and forms, such as status updates and comments; text boxes are limited in size (down to Twitter's famous 140 characters), image file size is capped at 100K, and so on. To be sure, the activities that take place in a Turing Test or via social media are complex, but they remain delimited. Thus, although we imagine the social stream as a collection of voices—*human* voices, *authentic* voices, *confessing* voices—flowing through our browser screen, in fact these are mediated chunks of data, pushed out from a server farm and packaged in protocological wrappers (HTML, XML, Javascript, CSS, and so on). We imagine the interactions to be real-time, but in fact there is a flow—albeit extremely fast—from client-user to server farm to client-user. These constraints, however cleverly coded, are the spaces into which a socialbot program can flow and possibly fool other humans, at least in a certain percentage of cases; just as the machine is hidden in Turing's game, so is the socialbot hidden in Facebook and Twitter.

"Success" in the world of socialbots is similar to success in a Turing Test: it is a game of probabilities and percentages. Reports from the software engineers who build them are replete with quantifications: "Over the course of the experimental period (21 days), socialbots were able to attract a total of 561 followers—an average of about 62 followers per socialbot."[64] In another, such social networks "as Facebook . . . can be infiltrated with a success rate of up to

80%."[65] Commercial-grade socialbots, such as the Twitter Bot sold online, promise to increase the user's number of followers by 1,500 to 3,500 per day.[66] These success rates are possible because the socialbot automatically and tirelessly sends a much higher number of friend or follower requests to the millions of people with Facebook or Twitter accounts. These bots play the odds.

By the standards of the Turing Test, are these socialbots intelligent? Have they fooled a large-enough percentage of human "judges" within social networks? These are tantalizing questions, but they are not my focus here. Instead, I want to suggest that the ability of socialbots to pass as human might be more a function of the a priori reduction of human activity to predetermined datasets than due to the coding skills of socialbot engineers. And with such reduction in place, the modulation of perspectives, thoughts, and communication among publics is a much simpler task.

Conclusion: Socialbots and Noopower

For Lazzarato, the nineteenth-century sociologist Tarde offers a way into thinking through noopolitics. Tarde conceives the public sphere "as a gigantic, instantaneous brain."[67] He sees social coordination as the action at a distance of one mind to another, evidenced by mass media communication systems. Mind-to-mind communication and instantaneous dispersion of inventions and knowledge are akin to neurons in a larger brain. This description is particularly resonant today as we think about the Internet (the "new home of Mind," to borrow a phrase from John Perry Barlow).[68] Regardless of the form of the communication network, Tarde's sociology holds that institutions that have "control of opinion, of language, of regimes of signs, of the circulation of knowledge, of consumption"[69] are the truly powerful entities in a mind-to-mind society. The institutions of noopower, then, are marketers, educators, and the sciences. In this sense, Tarde's work makes the move anticipated by Marx, who argues that the "general intellect" (the collective knowledge-power of society) would be the site of social struggle in the future. How can this general intellect be made (properly) productive? As Akseli

Virtanen argues, this is achieved via noopower over moods, sentiments, and habits of mind:

> The organization of immaterial production is possible only through the management of the general conditions of human action and communication, through organizing the general conditions of organizing. This organization of organization does not operate at the level of actual action or plain intimidation but on that of anxiety and inadequacy; not by confinement or demanding obedience to the rules and being afraid of their violation, but by setting expectations, moods, opinion climates, standards of communication and cooperation. It is the only way to control and organize labor power as an immaterial power, that is, not at the level of actual acts or products but on the level of potentiality and possibilities of life.[70]

Thus, rather than confine knowledge workers in factories or prisons, institutions of noopower make them productive by shaping the protocols and contours of their emotional and cognitive interactions.

Socialbots are one of many technologies developed over the past century to do this work of organizing and establishing moods and the parameters of cooperation. As a position paper for the aptly named Pacific Social Architecting Corporation puts it, "The vision of this technology is to enable operators to actively mold and shape the social topology of human networks online to produce desired outcomes."[71] Another set of software engineers explains, "In the future, social robots may be able to subtly shape and influence targets across much larger user networks, driving them to connect with (or disconnect from) targets, or to share opinions and shape consensus in a particular direction."[72] Examples the authors offer include benign goals, such as healing rifts among social groups and fighting misinformation, and certainly these can be goals of noopower institutions. But we can easily imagine more sinister uses for this technology—the example of the military's "persona-management" program points to one such possibility.

In many ways, the U.S. Air Force prompted the attention given to socialbot development by soliciting bids for persona-management software in 2011. According to the solicitation, this software

> will allow 10 personas per user, replete with background, history, supporting details, and cyber presences that are technically, culturally and geographically consistent. Individual applications will enable an operator to exercise a number of different online persons from the same workstation and without fear of being discovered by sophisticated adversaries. Personas must be able to appear to originate in nearly any part of the world and can interact through conventional online services and social media platforms. The service includes a user friendly application environment to maximize the user's situational awareness by displaying real-time local information.[73]

A contract for such software was awarded to a California company, Ntrepid. The goals are obvious: gather intelligence, build consensus, and influence opinions twenty-four hours a day via a network of socialbots. Here we can see the desire to encode culture into software to enhance the "soft power" of military leaders. And for state leaders, such bots can automate the process of quelling dissent. As Evgeny Morozov notes, "Following the Arab Spring uprisings, anyone posting critical comments about Bahrain or Syria on Twitter was likely to receive angry corrections from the government loyalists or, more likely, their bots."[74] Morozov notes that these actions were crude, but I would add that we are only witnessing the early versions of such online agents.

Moreover, given the history of marketing, marketers will use socialbots to "bring company mascots and assets to life with little effort," as one marketing professional puts it.[75] Brands and logos could move away from catchy slogans repeated ad nauseam ("They're g-r-r-r-eat!") to automated, personalized interactions ("Let's talk about what you did today after eating Frosted Flakes!") that are responsive to the emotional cues of the human interacting with it. Extending Arlie Hochschild's landmark analysis in *The Managed Heart*, if service workers have been trained to process customers and produce pleasant "states of mind,"[76] socialbots merely represent, and continue, this labor in software form. We can now imagine a modified Turing Test in which the jury must tell the human, the robot, and the ephemeral brand apart. Which one is most intelligent? Which one is most emotionally competent? Which one fools the jury the highest percentage of the time? If we learn that the branded socialbot has a mind and a soul, how terrifying would that be?[77]

Socialbots, thus, can be powerful tools to set "expectations, moods, opinion climates, [and] standards of communication and cooperation."[78] The existence of socialbots demonstrates that social media users are producing enough discrete states of mind to be imitated by Turing's Universal Machine. It should be clear that any system in which such user activities as expressions of opinion, desire, and emotion can be standardized and typified to the point that even robots can do the work of friending, liking, and relationship management is a boon to a highly rationalized system like social media capitalism. In the end, socialbots do not tell us much about *how* personal data are abstracted from users, only that data are being abstracted, and, moreover, that the data are standardized to the point that the information can be imitated by bots. The next few chapters explore the "how" of this process.

2

The Archive and the Processor

The Internal Hardware Logic of Social Media

> Forget the browser; real-time is the new crack.
>
> —Geert Lovink, *Networks without a Cause*

In 2008, during Mark Zuckerberg's first profile on CBS's *60 Minutes*, he helped reporter Lesley Stahl create her own Facebook profile.[1] He guided her through the template, even doing the work of typing in and selecting her "likes" for her. "Within a few minutes," Stahl reports, somewhat surprised, "I got a friend request" from someone she had not talked to in two years. Moments of inputting data into Facebook thus resulted in the elimination of years of lost time. Stahl notes that the near-instantaneous connection to friends is a reason why Facebook is so "addictive."

Speed, the new, and immediacy appear to be at the heart of Facebook, along with nearly every other social media site. If we take them at "interface value," such sites as Twitter, Google, YouTube, and Digg and such formats as the blog privilege newness over other forms of organization. These sites are exemplars of the "Web as platform," where the Web is treated as an operating system, and social media sites are applications built on top of the Web. Judging them by their interfaces, these "apps" are dedicated to social connection and instant access to information, much to the delight of users, such as Stahl.

And yet, pushing past the glossy, AJAX-driven interfaces of social media, we confront another element of this business practice. Social media sites are not simply surfaces dedicated to immediacy; they also comprise vast server farms with rooms of computers humming away. Of course, these servers provide some of the processing power that drives the immediacy of a social media site. But they also provide a function extremely necessary to any social media business plan: rationalized storage of vast amounts of data. In other words, while Stahl constructs her profile, Zuckerberg's servers are busily storing her data: her demographic information, the time she spends on the site, and her growing social connections (called the "social graph" by Zuckerberg). Here, we confront a contradiction: the smooth interfaces that users enjoy appear to solely comprise immediate connections and instant information, but the servers powering them are maintained in large part because of their long-term archival potential. This contradiction is the motor that drives social media.

If we open those servers, we see that the social media contradiction has its roots in the development of the modern computer itself, which is a synthesis of the immediate (in the form of the CPU or processor) and the archival (in the form of memory and storage of data). This fundamental architectural logic has informed the design of social media, and not just in terms of its technical facts. The social structures privileged by Facebook, Google, and Twitter also draw on the speed and storage dichotomy. The fundamental architecture of the computer is linked to the logic of social media, because a social dichotomy is at work on the basis of and reflecting (if not directly determined by) this architecture. In the hegemonic social media business model, users are encouraged to focus on the new and the immediate. They are expected to process digital objects by sharing content, making connections, ranking cultural artifacts, and producing new digital content, a mode of computing I call "affective processing." In essence, this business model imagines users to be potential superprocessors. With enough users aggregated via network effects and presented with a smooth interface (preferably something simple and binary, such as a "Like," "Tweet," "+1," or "Digg" button), they become a valuable source of digital-artifact processing.

In contrast, the archival possibilities of computers are typically commanded by social media site owners. They monitor users' every action, store the resulting data, protect the information via such artificial barriers as intellectual property, analyze it for clues about users' desires and habits, and sell the data for profit. This mode of new media capitalism prompts site designers to build websites that are capable of inscribing user activity into increasingly precise databases. Because of many sites' Terms of Service agreements, users cannot control these archives. These archives comprise the products of affective processing; they are archives of affect, sites of decontextualized data that can be rearranged by the site owners to construct particular forms of knowledge about social media users.

The impact of this sociotechnical dichotomy is tremendous. If Jacques Derrida,[2] Michel Foucault,[3] Marlene Manoff,[4] and Geoffrey Bowker[5] are right in arguing that control of the archive leads to social power, then social media site owners are becoming quite powerful indeed, because they have the ability to pull data from their archives to produce knowledge. New media capitalists seek to exchange these archives of affect with third parties (most commonly advertisers and marketers, a process covered in Chapter 4) to gain greater amounts of that classical storage unit of social power: monetary wealth.

To explore this contradiction between the immediate and the archived, I first outline the roots of the processor/storage dichotomy in the Von Neumann Architecture approach to computer design. I then examine how social media sites encourage users to value the new and to engage in the affective processing of digital artifacts. I include three brief case studies important in the history of social media: the NASA Clickworkers project, Digg, and the Amazon Mechanical Turk. Next, I explore how social media sites archive the products of user-generated affective processing. I draw on Karl Marx's Money-Commodity-Money' (M-C-M') circuit to illustrate how archiving user activities is a means to build social power. Finally, I conclude by examining the power of archives.

The Von Neumann Architecture

A basic architectural feature of computers is the separation of the processor and the archive. This architecture, dating to the mid-1940s and commonly referred to as the Von Neumann Architecture, calls for computer designers to store data and programs in a memory core and to process that data and execute those programs with the processor.[6] It was first described by mathematician John von Neumann and used on the prototype EDVAC, the first stored-program computer and the predecessor of all modern machines.[7] In this design, the storage unit of the machine and its processing unit are related to one another in a linear hierarchy of "fetch-execute," where the processor fetches data from storage, manipulates them, and then moves on to the next line of data. The processor focuses on only the immediate data it is working with, whereas the storage unit contains an archive of all the computer's command code and data.

The processor was built for speed and discrete operations. It manipulates small chunks of data as quickly as possible, moving sequentially through each element of complex equations. The faster the processor moves through each instruction, the faster it produces results for users. The processor is thus a mechanical/electrical replacement for the collected labor of large groups of human computers, a prior form of information processing used since the 1700s.[8] After all, "computer" used to mean "a person who computes"; this history of human labor informs the shape of our current machines. These groups of human computers were deployed in what Charles Babbage calls a division of "mental labor"[9] organized around mathematical operations; some would divide, some would multiply, and some would calculate square roots in a factory-inspired assembly line. Management of these laborers was achieved by using instruction books and hierarchical supervision. Stored-program computing pioneers von Neumann, John Mauchly, and J. Presper Eckert were intimately familiar with the human-computer process; all were involved with the Aberdeen Proving Ground, where ballistics calculations were conducted by human computers. As computer historian David Alan Grier puts it, "The architecture of these first [electronic] computers closely paralleled the structure of the human computing organizations. J. Presper Eckert Jr. once described the ENIAC as a 'ganged adding machine,' a phrase that could easily describe most of the [human] computing groups equally well."[10] Mauchly and Eckert's work on ENIAC was meant to supplement and possibly replace human computation. Similarly, von Neumann's description of the processor divides its functions along the logical lines of mathematical operations to increase its speed,[11] thus drawing on the division of labor already used in human computing.[12] This internal division of labor is now a standard feature of processors.[13]

Improvements in processor speed have altered user perceptions over the history of computing. By the 1960s, computer designers strove to make the computer seem as though it were reacting immediately to the whims of the individual user, even if it was being "time-shared" among multiple users. Paul E. Ceruzzi's history of modern computing singles out the Digital Electronic Corporation's (DEC's) PDP-1 and PDP-10 minicomputers as the earliest and most important exemplars of this design goal.[14] The PDP-1 established a new architecture at a lower cost than large mainframes, thus allowing more people to interact with it. Its descendant, the PDP-10, "was the system that first created the illusion of personal computing," setting it in contrast with other machines that required users to queue up, load their punch cards, and receive the processed data in a mode called "batch processing."[15] The PDP-1 and PDP-10 gave users the illusion of total control of the machine, as if the machine were reacting immediately to users' whims. The older model of batch processing made users consider computers to be exotic machines, sequestered in dark rooms and controlled by a priesthood of computer scientists who allowed outsiders only limited access. Ceruzzi argues that the immediacy of the PDP-1 and PDP-10 created a "mental model" of computing that has influenced our contemporary computing culture. For example, Microsoft founder Bill Gates spent his formative years working with a PDP-10; his disk operating systems (such as MS-DOS) and later Microsoft's GUI-based Windows operating system were in part attempts to recreate for users this sense of immediacy and instant control.

Despite efforts to replace the Von Neumann Architecture,[16] it has persisted as a fundamental design structure of computers. Regardless of whether it is most efficient model of computer architecture, computer users have come to expect immediate processing and interaction; this is our contemporary mental model of computing. For example, upon its release, Microsoft's Windows 7 operating system was subjected to performance tests by various computer magazines and websites. Popular perception of Microsoft's previous operating system, Windows Vista, was that it was slow and demanded too much of hardware. Windows 7 was under scrutiny for one main reason: was it faster? *PC World*'s answer is, yes, slightly.[17] *PC World*'s testing involved boot times, processing speeds, shutdown speeds, and the load time of applications (such as word processors and spreadsheets) on various computers.[18] This focus on speed is a very common one, growing out of increases in processor speeds that have in turn further generated an expectation of immediacy. Any new operating system is often judged on how well it utilizes the full power of the processor.

In contrast, memory is the archival potential of the computer. Its development is based on increasingly shifting data out of time. In the 1940s, while engineers strove to have the machine process data as fast as possible, in many cases (specifically those times that a remainder had to be carried over in a mathematical operation), some data had to be delayed momentarily. The memory developed by Eckert and used in the EDVAC was based on mercury delay lines. Like their name implies, these devices used the differences in the speed of sound traveling through different media to delay certain bits of information, transducing information through mercury and thus effectively storing it.[19] In the Von Neumann Architecture, this technique is used extensively to include not only short-term storage of numbers for operations but also long-term storage of computer programs. Computer memory is this time delay writ large. Data are taken out of time and stored as indefinitely as the medium will allow: a few seconds for the 1940s-era mercury delay line, to years, perhaps centuries, with disk drives and solid-state drives.[20] Again, this practice hearkens to the management of human-computer labor: the data used and produced by human computers as they did their discrete operations were stored for long-term use in such artifacts as books of artillery tables.[21] Long-term storage is always part of computation.

Like the processor, developments in computer memory shifted the ways in which users interpreted the machine. During the 1960s, the days of the mainframe, data were most often stored external to the machine on punch cards. These data were toted to the machine and loaded, and then, after the information was processed, the machine produced calculations. The development of tape reels and core memory marked a transition from such batch processing to Random Access Memory (RAM), a more efficient form of storage. With the advent of spinning disks and later solid-state drives, mass storage and access to data and instructions became possible. Mass storage was quickly adopted, because it made computers much easier to modify for different tasks, increased the amount of data the processor could work on, and allowed storage of documents and digital artifacts.

Thus, often when we talk about a computer, we discuss two contrasting facets: how fast can it process, and how much data can it store? These are the basic architectural facts of the technology, the result of design decisions made more than a half-century ago. The post–von Neumann computer is therefore a unique synthesis of immediacy and archival capacity.

This synthesis has been replicated on the Internet. The dual logic of the processor and the archive animates and in part determines the current business practice and social structures of the Web. The challenge for Web media companies is to always have new content to gain relevance in search engines and attract viewers. New media capital is meeting this challenge with the business practice of Web 2.0. According to technologist Tim O'Reilly, Web 2.0 is the practice of getting users to add value to a website by having them build its content, thus accelerating the cycle of media production so that sites become dynamic, constantly updated sources of new material.[22] Users of all abilities—from professional to semiprofessional to amateur—are asked to create videos, write blogs, post comments, and rank media objects. Such sites as Facebook, Myspace, Twitter, Google, Amazon, and Digg have enabled this constant production of content by ceding control over the immediate to users. They have essentially built empty templates (or as sumoti.iki calls them, "ghostly frames") and invited users to fill them in. Because of this practice, users now have unprecedented power over popular trends on the Web.

However, the catch here is that these site owners have not ceded the other half of the computational equation: the archive. While users are defining trends and shaping the now, social media site owners

are carefully shifting user-generated content out of time. Site owners command the past, a past largely imagined to be an increasingly
granular map of user desires. The architecture of social media not
only consists of empty templates; it also relies upon massive server
farms to store the content and associated data that users produce.

Interfaces of the New: "What Are You Doing Right Now?"

At the interface level, social media privileges the immediate. On Facebook, users are confronted with a little text box and a prompt asking,
"What's on your mind?" Twitter asks its users, "What's happening?"
Before its collapse in 2011, the social network Myspace asked, "What
are you doing right now?" These not-so-subtle prompts ask the users
of these social media sites to react, to present current "statuses": I'm
happy; I'm going to the airport; I'm texting my friends; I'm listening
to Radiohead.

This emphasis on the immediate is not limited to social networks
but also seen in media-sharing sites. Flickr's home page presents visitors with a count of photos uploaded "in the last minute." YouTube's
home page highlights "Videos being watched right now." Vimeo has
a tab on its main page for videos being shown "Right Now." Hulu
has a "Recently uploaded" page that features the latest video uploads.

Blogs and comment fields are also sites of immediacy. They typically read in reverse-chronological order; the newest post is on top,
with older posts pushed down the page. Likewise, comments fields on
many newspaper sites are organized in reverse-chronological order.
The old is pushed down; the new is always on top. The new is valued;
to follow the threads of discussion, one must click through pages of
comments and attempt to reconstruct a conversation back through
time. As Ben Elowitz argues, "The world changes fast now—and
readers have come to accept that the facts will too. Publishing rumors
and single-sourced stories (disclosed for what they are) is fair game for
winning audiences. . . . [T]he audience values timeliness more than
correctness."[23] Editors of blogs and news sites realize this, pushing
content far ahead of fact-checking and accuracy simply because it
reflects what is happening *right now*.

Of course, these sites are augmented by the developments of mobile computing and smartphones, which allow users to update their statuses, comment on digital artifacts, and upload content from wherever they can get onto their networks. Telcom company Sprint offers "The Now Network." In advertisements across the United States, the Verizon Guy used to famously ask, "Can you hear me now?" Users seek networks that can keep them connected wherever they are so that they can continue to engage with new information streams. This emphasis on speed is so compelling that mobile companies are increasingly using location-aware software to immediately alert users to consumer opportunities as they navigate spaces.

Finally, a new development in search is "real-time search" on such sites as SocialMention and Scoopler, which promise to return search results based on the newest streams on Twitter and Facebook. Google has responded with "Hot Trends," a list of the most popular recent searches. Bing's home page features "Popular Now" search terms. All search engines constantly send out spiders, robots that crawl the Web and report the latest changes on billions of websites. Much of this emphasis on the new is a result of Facebook and Twitter, two social networks that offer constantly updated streams of affect and varying degrees of public access to their latest trending topics.

In sum, as Chris Gerben notes, social media's user interfaces heavily emphasize the new and the immediate, even at the cost of other modes of organization, such as relevance or importance: "Digital texts not only privilege newness as a default design principle, but also rely on user-produced newness in order to maintain popularity."[24] Similarly, David Berry argues that network theory—a mode of inquiry often deployed by architects of social media—"privilege[s] a reading of reality that highlights the synchronic dispersal over the diachronic unfolding" and that "networks, in a certain sense, abolish history and shift our focus to the event, the happening or the now."[25] In their examination of Myspace, Mark Coté and Jennifer Pybus argue that users of social media sites are engaged in a "never-ending process of becoming. . . . Each new device and resource expands the capacity of their 'digital body' and allows them to forge new compositions of relations."[26] This dual reliance on user-generated "newness" and the emphasis on always becoming are built into the architecture of social media. It imagines subjects that are always connected, always updating, always searching, and never stopping their restless motion from one social network to the next.

However, this emphasis on the new is not, in fact, new. Rather, it is latest in the long-standing sociotechnological development of computer processing. As Adrian Mackenzie argued in the 1990s, the focus on the new was part of the two dialectical processes of the Internet: the emphasis on "real-time drives" and the archival impulse.[27] Using the language of virtual culture, he writes that "the virtual . . . can be positioned at the interactive threshold between the processes of real-time and the processes of the archive."[28] He rightly sees this dialectic in the structure of the computer discussed above: the Von Neumann Architecture of processor and memory. This dichotomy was built into the Internet from its earliest days. The emphasis on real time is thus a product of the Internet's architecture, which assumes that the end-user is interested in getting data fast. As Paul Virilio argues, *"The reality of information is entirely contained in its speed of dissemination. . . . [S]peed is information itself!"*[29] This emphasis is also based on the short-term goals of processing and the increasing speed of traffic on the Internet, while the emphasis on the archive is part of a longer historical process, which I discuss below.

This dual focus of the Internet is only accelerated today. Arising directly out of faster Internet connections and new suites of Web-programming technology, such as AJAX (Asynchronous Javascript and XML), one of social media's most salient features is that it is as responsive as desktop software. In fact, the marketing literature for social media services presents online software as a *replacement* for desktop software. AJAX is a codification of a new relationship between server and client computer, where only the most immediately needed data are served to the client. A well-designed website utilizing AJAX requests from the server only the information the user is currently interested in; the entire site does not have to reload. Thus, social media site designers seek to replicate and surpass the surface-level immediacy of the desktop operating system.[30] In this environment, as AJAX manual writer Anthony T. Holdener puts it, "The user will perceive everything about the web application as being self-contained. With this technology a savvy developer can make an application function in virtually the same way, whether on the Web or on the desktop."[31]

Web users are engaging with this immediacy and skimming along its surface by feeding updates into it and relying on it to provide emotional contact instantaneously. As Turkle argues, "We live a contradiction: Insisting that our world is increasingly complex, we nevertheless have created a communications culture that has decreased the time available for us to sit and think, uninterrupted. We are primed to receive a quick message to which we are expected to give a rapid response."[32] That is, the speed at which our electronic networks can connect us to others helps shape a new relationship to emotion: as Turkle explains, "Emotional life can move from 'I have a feeling, I want to call a friend,' to 'I want to feel something, I need to make a call.'"[33] The emphasis on the new in social media leads to immediate affective exchanges: I message you, and you chat with me. If you do not, I become anxious. Why are you not e-mailing me back *right now*? If, as Clay Shirky argues, our mediascape is marked by "filter failure"—that is, if we are unable to filter all the possible content we might encounter[34]—then perhaps this emphasis on the new is logical. A connection (however weak) with a friend *right now* might outweigh the value of terabytes of uncontextualized data that may or may not offer emotional or social value.

However, this is not just a structure determined by the computer's technological architecture or by the users' actions and desires; it is also determined and extended by the needs of the heterogeneous engineers of social media. These engineers are operating within a particular milieu: late capitalism. Ben Agger has aptly named this formation "fast capitalism," arguing that it has radically extended Frederick Taylor's vision of scientific management.[35] Virilio's arguments about dromology (time-space compression) orient us to the use of fast-capitalist tools, such as instant communications and instantaneous navigation of digital spaces.[36] Digital environments condition users to expect information immediately and thus to act on it. In some senses, this is a social good; it enables us to access greater spheres of information than our ancestors could. But in many ways, the anxiety-producing speed of social media is part and parcel of a larger emphasis on knowledge work in late capitalism. As Tiziana Terranova argues, "The Internet is about the extraction of value out of continuous, updateable work, and it is extremely labor intensive. It is not enough to produce a good Web site, you need to update it continuously to maintain interest in it and fight off obsolescence."[37] It is as though we fight entropy with entropy in an ever-losing battle.

Thus, this phenomenon has not arisen in a social vacuum. When we consider this focus on the new as another instance of the just-in-time demand for labor that marks late capitalism, particularly (but most certainly not limited to) affective immaterial labor, this emphasis on the new is clearly a case of media and Web corporations relying on users to do the work of processing digital artifacts and personal data and generate an emotional surplus. Users are relied on contingently and intermittently, but relied on nonetheless. In short, while users have become accustomed to instantaneous action from their networked devices and instantaneous connections to their friends, capitalists, investors, and media companies have become accustomed to users' near-instantaneous processing of data and have positioned themselves to exploit and, as we shall see, archive the results of this processing.

Crowdsourcing: From Mars to Digg to the Mechanical Turk

To illustrate this, I offer three examples that loom large in the mythology of social media. The first is a nonprofit volunteer effort. In 2000, NASA began its Clickworkers project, a small, part-time project that allowed public volunteers to mark craters in photographs of Mars. Marking craters is a tedious and time-consuming task for an individual; according to Michael Szpir, "The task is usually undertaken by someone trained in the art and science of rating craters, but there are many thousands of craters on the planet and, well, most scientists (even graduate students) have better things to do."[38] Seeking a more efficient way, the Clickworkers project was an experiment to find out whether public volunteers could process those images faster and more reliably than the handful of scientists who would have done the work. It was a resounding success. According to Yochai Benkler, more than eighty-five thousand volunteers visited the site and created more than 1.9 million entries: "An analysis of the quality of markings showed 'that the automatically computed consensus of a large number of clickworkers is virtually indistinguishable from the inputs of a geologist with years of experience in identifying Mars craters.'"[39] These contributions were done by volunteers, many of whom spent a mere five minutes on the site before moving on. As a part-time experiment, the project was staffed by one engineer with two consulting scientists. As such, it created a tremendous savings in time and resources for NASA, and it continues to this day.[40] But, more importantly, it demonstrated that the Internet provides a structure for massively distributed human processing; users from all over the world lent a few minutes of their visual acuity to the project, and these micromoments of labor and attention aggregated into an incredible superprocessor.[41]

The Clickworkers project has a central, almost mythological place in the arguments of social media enthusiasts, including Benkler[42] and Jeff Howe[43] (who coined the term "crowdsourcing"). For Benkler and Howe, the production of knowledge has finally been "democratized," broken out of the confines of expertise, space, and certification. The Clickworkers project proves that users will volunteer to help an institution (in this case, NASA) achieve a goal. Users' online activities are presented as "spare computing cycles,"[44] likened to the spare processing cycles of an idling computer. For Benkler and Howe, this also means that corporations must take advantage of users' free labor, since to choose otherwise would be to make an irrational business decision. Distributed, networked labor, they argue, is now much cheaper for capital to rely on. This iteration of capitalism, dubbed by Benkler the "networked information economy," involves "decentralized individual action—specifically, new and important cooperative and coordinate action carried out through radically distributed, nonmarket mechanisms that do not depend on proprietary strategies."[45] While he uses the term "nonmarket," it is clear that Benkler as well as Howe see this development as, in fact, a new, cheap labor market.

As O'Reilly argued in 2004, corporations have responded to this phenomenon, recognizing the potential value created by masses of users doing small tasks within a website.[46] Social networks, video-sharing sites, auction sites, and search engines rely on the labor of users to create their content. Here, I want to focus on two examples of for-profit distributed human computing.

Digg

As discussed above, media scholar Shirky has argued that users of the Web (and other media) suffer from "filter failure"; they are simply drowning in media objects and are unable to decide which objects are relevant.[47] In the history of social media, Digg was an important response to filter failure. Prior to its sale and redesign in 2012,[48] the service offered users a way to sift through the mass of digital material on the Web. This was accomplished by the work of users who did one or more of three tasks: submitted material, rated it (a process called "digging" or "burying"), or commented on it. If an item received enough positive "diggs," it would reach the front page, where millions of visitors could see it and link to it. Conversely, items could get buried by Digg users, either because they were seen as irrelevant, not entertaining, or spam. In addition, the submissions were further sorted by users, who categorized them into subsections, such as Technology,[49] World and Business, and Gaming, each with its own sub-subsections. In this way, users could collectively sort and rate the vast material available on the Web, presenting a structured snapshot of what is popular online.

Digg cofounder Kevin Rose has argued that a system of user-led Web curation returns power to "the masses": "This was the first time that anyone experimented with allowing the general mass audience to decide what they believed to be the most important topic of the day."[50] Indeed, in many descriptions of the site, it is as if there are no administrators, investors, or site owners at all; as *How Stuff Works* writer Julia Layton described Digg in 2006, the only agents involved in the site were varying grades of users, from casual to "dedicated."[51] Even her description of the server-client structure of the site—a complex arrangement of hardware and software that required IT labor to run it—elides any other persons laboring on the site. Users were ostensibly in control, and it showed: between 2004 and 2012 (when the site was sold and later rebuilt), traffic increased, and the number of stories linked on Digg rose almost exponentially.

Rose's "masses" could confront this material anonymously at Digg.com, or they could sign up for accounts that promised to further refine what they saw on the basis of their tastes. Account holders could enjoy all the benefits of the social Web, connecting with friends and having stories automatically suggested to them (features that are now even more emphasized in the post-2012 rebranding of the site). Of course, Digg carefully observed and stored the digitized activities of these users, a process I expand on later.

For a period during the late 2000s when Digg was at the height of its popularity, the results of this sorting, ranking, and surveillance were distributed across the Web in widgets that proclaimed that the news items they contained were "Powered by Digg's Users," a direct homage to Intel's famous "Powered by Intel" stickers. Digg thus explicitly compared its user base to a microprocessor, no doubt implying that the millions of Diggers who sorted and ranked items were more powerful than any software algorithm. Judging the effectiveness of Digg versus a computer algorithm–based model (for example, news. Google.com) is beyond the scope of this chapter. However, what Digg did offer (and many similar social bookmarking and rating sites, such as Reddit, still offer) was an easily understood numerical assessment of its highly rated items. Moreover, it offered the essential Web 2.0 ingredients of speed and the new. As Digg cofounder Jay Adelson puts it:

> [Digg] attracted the attention of the news media immediately—the fact that we had this incredible speed. Automated systems take time to crawl the net. Editorial systems have the human factor. They may decide they're not interested that day, or they'll do it tomorrow. In our case, there's no barrier, so the second a story would be interesting to this mass public, we can break it.[52]

This emphasis on the new in news appeals to those Web users who seek immediate access to information. Without such a filter, this argument goes, users might miss out on news stories because they are navigating serendipitous content in potentially outdated sources, such as newspapers.

Like many social media sites, Digg has been described as democratic and antiauthoritarian, breaking down the distinction between editors, writers, and readers. However, it also came under criticism, because it has been revealed that many media and marketing companies paid for favorable publicity. For example, the website USocial.net attempted to sell diggs (as well as votes on other social news sites).[53] For a small fee, USocial offered two hundred votes on these sites; this number of votes could easily promote items to much-desired positions on Digg's front page. In hindsight, it would seem obvious that payola-type schemes would begin to appear on Digg. In fact, pre-2012 Digg had the potential to be an unregulated labor market, where Diggers could be hired by media companies and advertisers who wanted to promote their products. This practice was against Digg policy but often happened nonetheless. To be fair, Digg's original owners preferred the purity of unpaid Diggers and sought ways to direct their attention to legitimated submissions. Moreover, Digg relied on users to find and ferret out spam; Rose estimates that 95 percent of spam on the site was removed by users.[54]

But this desire for pure, unpaid digging reveals another ambition: the application of the Digg user–processed model to advertisements. Digg encouraged users to rate the advertising on Digg pages, a feature called "Digg Ads." As the Digg FAQ page explained:

> The goal of Digg Ads is to encourage advertisers to create content that is as interesting as organic Digg content. By Digging or burying the Digg Ads, you are helping us determine which ads to show to more people, and which ads to show less frequently. . . . Digging or burying Digg Ads helps us continue to improve the overall Digg experience. We give each advertiser a content score based in part on the community's Diggs and buries. Advertisers with higher content scores will pay less and their ads will be shown to more people. Also, when

you bury an ad you won't see the ad again (as long as you're logged in).[55]

Obviously, a pool of users who were being paid by third parties disrupted this "organic content." Ultimately, Digg sought to elide the nature of online marketing by integrating it into Digg's very fabric, a process not unlike product placement in movies and television.

In sum, Digg was an important site in the history of social media and the intersection between processing and archiving I am exploring here. It was built on the model that the NASA Clickworkers project pioneered: a distributed processor comprising a class of users clicking their way through news stories. Much like the Clickworkers project, Diggers did not need to spend more than a brief moment "digging" a story; the aggregation of these micromoments of labor added up to the Digg front page, a reduction of spam, a grading of advertisements, and a robust collection of commentary.

Amazon Mechanical Turk

Between 2004 and 2012, while Digg teetered on the edge of becoming a microlabor market where Diggers could potentially be hired to promote marketing materials and particular media objects, one social media site unabashedly and explicitly became such a labor market. The most important example of crowdsourcing as a new labor market is the Amazon Mechanical Turk. Publicly launched in 2005, the Mechanical Turk is a marketplace of tasks, which Amazon calls "Human Intelligence Tasks" (HITs). As the name implies, HITs emphasize tasks that require human judgment, such as image recognition or audio transcription. In essence, "Turkers" who complete HITs are marketed to employers as the world's best computer, combining the unparalleled capacity of humans who can read, recognize images, and make judgments with the immediacy of computers. As Jeff Barr and Luis Felipe Cabrera explain, Amazon envisioned the service as an answer to companies that need metadata improvement, image selection, and translation to be done on increasingly large scales.[56]

Computers cannot handle these types of tasks with any accuracy, and hiring workers to do them in-house is extremely expensive. Thus, the Mechanical Turk has utilized the Web as a means to connect companies in need of repetitive digital tasks to a worldwide labor market of microlaborers.

Despite its reliance on human processors, the Amazon Mechanical Turk's marketing literature abstracts the *human* processing that takes place during HITs. Amazon wryly calls this "artificial artificial intelligence," referencing the interface, which makes human work look mechanical. It is structured like the server-client practice of networked computing: the employer sends a request to Amazon, and the humans' response to the request is returned via AJAX-style programming. The legendary marketplace, where labor meets capital in a personified negotiation, is replaced by a screen interface, where labor finally becomes completely mechanical and rationalized. Human labor is reduced to cost, a mere input in the production process, and a cheap one at that. In many cases, HITs are worth a few (U.S.) cents a task. For example, as of this writing, one HIT asks Turkers to classify advertisements for 5 cents in three minutes. Another asks Turkers to "check if these websites work" for one penny a piece.

Like Digg, the Mechanical Turk is built on the Clickworker model, but it takes that model further by emphasizing the processing of digital artifacts and deemphasizing knowledge of what these tasks are for. Turkers are encouraged to ignore everything but the microlabor task at hand. While humans-as-laborers are elided in the structure of Mechanical Turk, employers are also hidden behind layers of abstraction. As Jonathan Zittrain explains, Turkers do not have much knowledge (if any) of their employers—they simply have Amazon accounts and receive micropayments for services rendered. Zittrain argues that this system could potentially be put to nefarious use; he imagines the government of Iran creating HITs that sort images of Iranian citizens into two categories: protester or potential informant. This HIT could easily be structured to cover up the identity of the employer, while the Turkers involved would blithely sort photographs for pennies apiece.[57] Of course, this is an extreme example, but it points to the highly abstracted nature of this site; users here are imagined as processors, meant to do tasks quickly and accurately and then return the results to unseen entities.

While the social media emphasis on collective intelligence and the wisdom of crowds is compelling, the goal of commercial social media sites is to capture the processing power (and subsequent value) of a critical mass of users, either directly (as in the case of Digg) or indirectly (as in the case of the Amazon Mechanical Turk). The owners of the sites do not particularly care *what* the users are processing, so long as their attention is fixed on the site. In short, the development of social media out of this history is a trajectory of increasing capitalization of the processing power of the masses of computer users. Whereas computer scientists and software engineers might have dreamed of building truly universal machines, ones that could fully replace humans, in many ways computers still cannot rival a mass of humans. No computer can compete with us when we join and tackle problems. But what do we do with this capacity? In social media, what began as an ethic of nonprofit volunteering to a greater cause (NASA Clickworkers)[58] has morphed to an individualistic emphasis on sharing and personal connection (Facebook, Google+, Twitter, YouTube) and even to the ultimate just-in-time flexible labor market (Amazon Mechanical Turk). This emphasis is reinforced by the predominant focus on the new. The user has to update his or her status, check on friends, make new friends, and check in again for a new connection or emotion, while Turkers seek the latest HIT. In this milieu, computer users are imagined to be the processors that computers never could be. However, computers do have humans trumped in another area: memory.

Engineering an Archive of Culture and Affect

While computer scientists could not replace human skills, such as image recognition and subjective ranking, with artificial intelligence, the other half of the computer's architecture has been much easier to construct, expand, and improve on. Memory is as essential to modern, von Neumann–inspired computers as is the processor. The processor works on data, but data (in the form of instructions and results) must be stored somewhere. Memory capacity has grown tremendously, leading to today's terabyte drives that store vast amounts of information.

This information must be routed to the processor. While contemporary computers share the same basic Von Neumann Architecture, the *connection* between memory and processing has become more and more complex. Busses, short-term caches of memory, and dedicated distinct pathways for instructions and data are some solutions to what has been called the "Von Neumann Bottleneck" between memory and processor.[59] A major focus of computer science has been to widen and improve the speed of these pathways (if not circumvent them altogether somehow).[60] Microprocessor developers, such as Intel, have seen their best efforts somewhat thwarted by the slower pace of development of memory speeds; memory technologies cannot keep up with processing speed.

And yet, this is not to say that memory is the ne'er-do-well little brother of computer architecture. While memory speeds are slow, and while the bottleneck remains a structural shortcoming, memory capacity has exploded in the past decade. According to UC-Berkeley's How Much Information? project, more than 5 exabytes (10^{18} bytes) of print, film, magnetic, and optical information were produced in 2002, with 92 percent of them stored on magnetic drives. This is thirty-seven thousand times as much information as is stored in the Library of Congress's seventeen million books.[61] Even in the time since the How Much Information? report, estimates now place the amount of information stored worldwide at 500 exabytes, ten times the 2002 level.[62] This storage has been made possible by advances in computer memory hardware.

Thus, we have a basic computer architecture: processor, memory, and the path between the two. What is germane to this chapter is not computer scientists' efforts to overcome the Von Neumann Bottleneck *inside* the computer, but their attempts to address a bottleneck between the human processing I outlined above and the vast pools of digital data in the world. Here, I want to answer these questions: if there is a glut of digital data stored in the memory banks of servers worldwide, and if distributed human processing is a free (or cheap) and efficient way of processing it, then how are humans and digital data interfaced?

The answer, of course, is the advent of widely distributed broadband Internet connections. Whereas dial-up connections had to be established by dialing a number and connecting, a process that can be unreliable and at the very least ties up phone lines in many people's homes, broadband connections, such as DSL, cable, and WiMax, are always on. The connection becomes silent (i.e., no more chirping sounds over phone lines) and invisible, since it does not get in the way of the user's online experience. Broadband connections are sine qua nons for social media. Without them, AJAX-based applications that replicate desktop software would not be viable. With them, users are increasingly engaging in constant connections to the Internet and to new forms of Web-based software. Moreover, since this constant connection is far more reliable than dial-up, it is akin to the dedicated busses installed between memory and processors within the Von Neumann Architecture. With these busses, site creators can imagine masses of users who will interact with digital material without worrying about the connection. Thus, such sites as Amazon Mechanical Turk can rely on users who are focused only on completing microtasks.

In addition, broadband enables not only the distributed human processing that the Clickworkers project, Digg, or Amazon requires but also the *storage* of the results of human processing. The data that humans process must be stored somewhere. This is an often overlooked aspect of always-on broadband connections. While broadband is very often presented to consumers as a fast way to *download* material, every download also requires *uploads*. At the very least, a client must send a request, such as the XMLHttpRequest object,[63] to a server to receive data. Thus even a user who "free rides" on a site, only browsing but not contributing content, generates such data.[64] These requests can easily be stored by the server, forming an archive of user activities that can be later analyzed and data-mined. Moreover, Web users rely on broadband connections to upload photos, movies, or blog posts. In short, just as data are necessarily and automatically migrated from memory to the processor and back in the Von Neumann Architecture of computing, archiving the results of user activities in social media is a built-in process. Capturing user activities in matrices of client-side request logs, XML metadata, and IP address logs is a *necessary* aspect of the broadband/AJAX connection between client and server. As users surf the surface of social media, the online archive grows ever more precise.

A-P-A'

Ultimately, for the owners of social media sites, the goal is to store as much user-generated content and data as possible, serve it to users who process it further, and then store the results, creating an ever-more-precise and extensive archive. Facebook is a prime example of this principle. To continue growing, it requires more participants to attract other participants. This is the so-called network effect, where a networked technology's value increases as more people use it. The network effect is apparent in any communications network; telephones, for example, are useful only if there is someone to call. Digg, Facebook, YouTube, and other social media take this a step further by expanding possible uses. On these sites, users do not simply e-mail one another, but play games, chat, give gifts (real or virtual), comment, post media objects, and display their statuses. These interactions are often (but not always) asynchronous. Users often interact not with one another in real time but rather with digital ephemera that stand in for users: avatars, status updates, images, and videos. Thus, what social media site users are interacting with is an archive of affect, digital objects that have meaning within the context of social connections. They are processing this digital archive: sorting their contacts into lists, liking this status update, commenting on that photograph, or sharing a virtual gift.

In this sense, the archival half of social media is a direct descendant of the transduction of information out of time, first achieved in mercury delay line memory in the EDVAC (as described above). In this case, the time-shift of digital data, such avatars, status updates, and comments, is extremely fast, but it is there nonetheless. After the user logs onto the social media site, his or her keystrokes and clicks flow to a central server farm and are stored, and then duplicates are transmitted back to the user's screen and the screens of his or her friends. The time between the keystroke and the return is fast, but there is always a delay. Moreover, this transduction, this shifting out of time, reflects the dialectic between use-value and exchange-value within commodities: a commodity's use-value is suspended in time until the exchange is completed.[65] Think of an iPad on an Apple Store's shelf: its value-in-use remains out of touch, excited in the consumer's imagination (excited, of course, by hype and advertising). It is only when the exchange is made that the use-value is realized (and of course, then, the surplus value locked in the commodity is also realized into money). In the case of social media, the use-value of access to the social network and connection to friends remains imaginary until after the exchange (logging in and agreeing to the Terms of Service—that is, the exchange of private data for access to the network) is completed. Here, despite the speed and immediacy of social media, we can also see a dialectic of timelessness as digital objects become suspended in time and stored for later analysis and, as the owners of social media sites hope, conversion into monetary wealth, the store of social power.[66]

We can think of this with a simple formula. Facebook seeks to have a large archive (A) of these objects for users to interact with. Facebook was initially seeded with such applications as the wall (an area for user comments), photo sharing, and notes. These basic applications allowed users to post text, photos, and comments on other users' profiles. As users interact with these objects, processing (P) them, Facebook watches their actions and collects data, archiving (A') this newly generated data. This is the information Facebook seeks to sell to advertisers. The process has been accelerated as Facebook has opened its Application Programming Interface (API) to third-party developers who create more applications inside which users interact. In sum, Facebook—and other social media sites—seek to grow the archive through the process A-P-A'. The larger the archive and the more granular the data about the desires, habits, and needs of users, the more valuable the archive. And if the archive is reliably linked to users who can sort data and process digital artifacts, the archive can be grown and made more precise.

Each of these steps is highly necessary, but only one can cause the archive to grow. As in the Marxian Money-Commodity-Money' formula that this formula echoes, the process that grows the archive is labor—in this case, the microlabor of users whom Ursula Huws would call "cybertariats,"[67] since the work in question is often highly casualized and even presented as entertainment. This is part of the larger exploitation of previously untapped "people power" on the Web, where the leisure of Web users who seek entertainment and diversion is finally made productive for globalized capitalism.[68] This is done through the explicit connection between computationalism and leisure. As Mark Zuckerberg explains:

> If you had asked people . . . ten years ago, "What would be the best way to take a big set of photos and identify the people in them?" most people would have said, "You should have a face recognition algorithm," or something like that, and kinda crunch all the photos. But it turns out that it is a lot easier to let people tag the photos of their friends, and create a good interface where you have your friends list there and you can do that. It works really well. . . . One person does manual work, and they do the work for the network.[69]

Whether they are digging, turking, tagging photos, or simply updating their statuses, users are explicitly imagined to be the labor/processor core that "powers" social media. They are the "Intel Inside" of social media. In sum, they are the processor in this reimagined Von Neumann Architecture, a social reflection of the internal division of labor that constitutes computers.

Conclusion: The Power of Archives

One of the major tropes of social media is that websites organized around users making decisions eliminate authority. Social media, this argument goes, remove gatekeepers, allowing average users to produce, evaluate, and distribute content. The future, as Axel Bruns proclaims, is "user-led," no longer the domain of executives who plan broadcast schedules and distribute media from centralized studios.[70] We are no longer beholden to the tyranny of mass media, argues Chris Anderson; now we can find whatever form of entertainment we desire in affective niches located somewhere on the "long tail" of decentralized, participatory content creation.[71]

However, authority is not truly eliminated in this new media environment. While social media may have, in fact, created new ways for users to find and manipulate digital content, the archival capacity of social media allows for new centralizations of power, hidden away beneath the abstractions of the smooth, user-friendly interfaces. Although traditional mass media gatekeeping roles may have been eroded, social media have enabled new media companies and entrepreneurs to assume curatorial roles; these curators build archives out of the products and traces of users' affective processing, protect them via Terms of Service agreements and intellectual property regimes, and mine them for profit.[72]

For example, the Facebook Terms of Service states:

> You own all of the content and information you post on Facebook, and you can control how it is shared through your privacy and application settings. For content that is covered by intellectual property rights . . . you specifically give us the following permission, subject to your privacy and application settings: you grant us a non-exclusive, transferable, sub-licensable, royalty-free, worldwide license to use any IP content that you post on or in connection with Facebook ("IP License"). This IP License ends when you delete your IP content or your account *unless your content has been shared with others, and they have not deleted it.*[73]

All is equitable until the last clause of the last line. Facebook claims no ownership over user intellectual property (assuming the user sets the privacy controls correctly). Facebook even will relinquish any claims to its licensed use of user material after account closure— *unless it has been shared.* Since the explicit purpose of Facebook is to allow users to share their photos and writings, Facebook cleverly captures user data in a perpetual license while denying its intention to do so. Unless the user's "friends" also delete the shared data, the information will *always* be licensed to Facebook. Facebook is a service that allows users to share among their "social graphs," but this is itself simultaneously an expression of a second, less-explicit purpose of the site: you may share with others while we capture the digital objects you share to gather data on your preferences and desires.[74]

But datasets are not in themselves archives. To be an archive, the material collected must be organized in a manner that allows for the post hoc construction of power/knowledge: "Indeed, how could one start constructing an archive without knowing the principle of its construction, without knowing in advance, among other things, what to select?"[75] The material collected must be done in anticipation of its future reconstruction. Briankle Chang sees the archive as existing in the future perfect: "They will have become what they already were."[76] This becoming echoes the self-production of users: they rely on social media to become what they already were. Their becoming and the data they produce are always already waiting for the archon (authority, curator) to appear as predicted in the future perfect. As Alan Sekula argues, "Clearly archives are not neutral; they embody the power inherent in accumulation, collection, and hoarding as well as the power inherent in the command of the lexicon and rules of language."[77] Geoffrey Bowker puts it very clearly: "What is stored in the archive is not facts, but disaggregated classifications that can at will be reassembled to take the form of facts about the world."[78] Thus, what is required is an authority to construct "facts" from the fragments that sit on the archive's shelves. Bowker's name for our computer-driven memory episteme is "potential memory," a mode of power where those with access to the archive create narratives post hoc from a priori taxonomically organized objects that are scattered across many physical storage sites.

Social media sites lend themselves to such post hoc constructions. Marketers, lawyers, entrepreneurs, social scientists, psychologists, and experts in so-called big data have built social media archives to construct exchangeable images of user/consumers. The "facts" that are then produced are largely concerned with consumer preferences.

Whereas state-based interpellation of identities might arise from the metrics of security and biopolitics (date of birth, race, country of origin), rationalized identities in social media arise from the metrics of capital and consumption: user profiles, categorized social connections ("friends," "coworkers," "family"), credit scores, searches, purchase histories, media consumption, desires, fantasies, demographics, and movements through space; that is, this is Deleuze's "'dividuation'" in action.[79] As far as marketers and investors are concerned, these are the most salient digital fragments to be stored in the servers of these sites.[80] However engaged users are with their tweets, profiles, articles, videos, and images, in this adaptation of the von Neumann division of computational labor, users are often reduced to affective processors working for the owners of the digital archive.

Thus, although some popular and academic accounts of social media often present them as eliminating editorial authority, by considering social media as an expression of the relationship between users/affective processors and the owners of digital archives, we can readily see that authority is alive and well online, transcending the Internet into neo-Hobbesian sovereigns that Jarod Lanier calls "the lords of cloud computing" who command data flows and storage.[81] Although editors and gatekeepers have seen their roles eroded, data miners have emerged as the new personification of media power. As Vincent Mosco argues, in the history of media technology in capitalism, power always reasserts itself in some form, despite the utopian proclamations of democracy and equality that accompany a new media form.[82] Here, the new shape of power is the power to freely sort, divide, relate, and manipulate data.[83]

Emphasizing the cultural, juridical, and economic power of archives adds complexity to the insights of such media scholars as Mark Andrejevic and Zittrain, who argue that the most salient effect of social media is a radical increase in surveillance in the digital enclosure. For Andrejevic, surveillance is presented by media companies as the "guarantor of individualism and self-expression and thereby as a means of overcoming the homogeneity of mass society."[84] In exchange for this guarantee of individuality, users agree to be watched as they interact, shop, and surf on the Web. Similarly, Zittrain argues that popular (or perhaps more likely hyped) fear of Internet viruses, identity theft, and cybercrime has driven consumers to embrace "closed" technologies, such as TiVos, smartphones, and operating systems with heavy-handed Digital Rights Management (DRM) software.[85] This "closed" architecture is easily surveyed by media companies and advertisers. Both scholars rely on the political economic language of enclosure to describe this metaphorical space in which users are being watched.

But of course, social media involve more than simple monitoring. We might live in the synopticon, where each of us watches one another and where capital surveys all, but this regime could not function without storing data and mining it after the fact. To use an analogy, a surveillance camera does not just watch people; it records their activities. If a crime is committed, the recording becomes evidence, but only after an authority watches the recording and pulls that material out of the archive. Moreover, it must be carefully handled, prepared, and contextualized for presentation in court.[86] Likewise, surveillance in the digital enclosure requires storage and retrieval. These activities become the basis for an archive of affect, from which Digg can pull particular bits of data out to arrange into digital images of users' desires for sale to marketers. Power thus arises from the ability to (1) close off this database from the rest of the Web and (2) pull disaggregated data from it and reconstruct the information into "facts" about users. As those users continue to process digital objects, the resolution of the digital images of user desires increases, providing clearer, albeit cleverly cropped, pictures of users.

All of this is not to say that users are disinterested in storage and creating archives. As I argue in my study of YouTube, users are engaged in archiving as well as processing, but their archival activities are more akin to the pattern associated with free laborers in Marx's C-M-C' circuit.[87] In that circuit, the free laborer (free in the sense that the laborer owns nothing but his or her own labor) meets with capital in the market and sells labor-power (C) in exchange for a wage (M), which is spent on consumable goods to sustain the laborer (C'). For the user of social media sites, the process might be described as P-A-P'. That is, the user processes digital objects to archive them for him- or herself and then (given the attention paid to the new) seeks new objects to process. The archival impulse is personal, immediate, driven by a cultural equation of digitization to memory. The archive for the user might contain lists of friends, pictures at a party, videos stored in YouTube, blog posts. However, users are less able to *reconstruct* the material in the archive into new facts about the world, because the scale of their archives is much smaller than the massive, server farm–enabled archives of centralized media companies.

Moreover, when any user tries to divest from the social Web by downloading personal digital artifacts, he or she is confronted by radical decontextualization. For example, consider Archive Facebook, a Firefox add-on that "allows you to save content from your Facebook account directly to your hard drive. Archive your photos, messages, activity stream, friends list, notes, events and groups."[88] Users who install this add-on in the hopes that they will be able to download all the data associated with their profiles will be disappointed: data produced by friends are not included. Legally, this is understandable; I should not be able to access my friends' data without their permission. But this limitation points to the gap between any personal archive and the massive archives maintained on social media sites: without the context provided by his or her social network, a user's data are atomized, floating freely without the connections made within the network. Removed from the context of the social media archive, personal data become stunted and incomplete, radically reduced in value. The user archive is a *petit-archive*.

Finally, new media companies are constantly seeking ways to reduce users' perceived need for a personal archive; instead, everything is to live in "the cloud." Media corporations are actively building read-only websites, such as Hulu and Netflix, which provide subscribers with instant access to television and movies, provided those users do not "steal" the content by storing it locally. This is the media industry's reaction to digital piracy on torrent networks. According to this logic, if media companies can provide legal and convenient access to digital content, torrent use will decline as users flock to centralized servers. The archive would shift from being radically distributed (stored locally on users' computers) to being centralized and hierarchical. Indeed, as Matthew Kirschenbaum shows, the hard drive and its storage capacity have gone from being novel and spectacular (in the form of Professor RAMAC at the 1958 World's Fair) to being encased and black-boxed (in the form of the hermetically sealed hard drive inside a desktop computer) to fading away into the clouds (in the form of networked storage). The push is for all user data to live online instead of locally.[89]

Thus, the gap between social media site owners and users is clear: social media site owners have access to all the data within their walls, and the users have legal access to only their own data, and even then access might be limited. This discrepancy arises because of the classic practice of private property; intellectual property rules allow new media capital to maintain its archives and severely curtail access to them—unless, of course, the user is willing to pay or provide personal data. The questions of privacy that arise from this surveillance are nearly always resolved at the individual level, but the breadth of social media archives demands that we take a wider view of how social media operate within the architecture of the processor and the archive.

In Chapter 3, I draw on software engineering practices to expand on the gap between users and site owners implied by the Von Neumann Architecture. As I show, the distance between ordinary users and site owners is expressed in code as much as it is in hardware.

3

Architecture and Implementation

Engineering Real (Software)
Abstractions in Social Media

> Last night I held my transistor radio in my hands, gently,
> as if it were alive. I examined it closely, searching for some
> flaw, some obvious damage. But there was nothing, no
> imperfection I could see. If there was something wrong, it
> was not evident by the smooth, hard plastic of the outside.
> All the mistakes would be on the inside, where you
> couldn't see, couldn't reach.
>
> —SHERMAN ALEXIE, *The Lone Ranger and Tonto Fistfight in Heaven*

On January 12, 2011, *Bloomberg News* publicly broke the announcement that Myspace CEO Mike Jones made to his employees: the site was either going to be sold or spun off from its parent company, News Corp.[1] The news came as little surprise, as Web-industry writers had been reporting on the demise of Myspace for at least two years. In 2009, Myspace laid off 30 percent of its employees, cutting four hundred jobs.[2] This was followed by a further cut two years later, reducing its workforce to roughly five hundred.[3] This restructuring of the company was a reflection of the downward slide of the site's traffic and revenues,[4] and Jones's announcement marked a low moment in Myspace's eight-year run. In late June 2011, Myspace was unceremoniously sold for $35 million to an advertising network.[5] Five years prior to the sale, Myspace was the most popular social networking site in the world and by some estimates the most visited website in America, beating Yahoo! and Google.[6] By the end of 2010, however, it was clear that Facebook was the dominant social networking service, and Myspace began rebranding itself as a "social entertainment" site, effectively ceding the social networking market to its rival Facebook.[7]

Myspace even went as far as to integrate many of its functions into the growing Facebook Connect service, which allows third-party sites to connect to Facebook users and offer customized interfaces based on user preferences.[8]

It is difficult to mourn the creative destruction of Myspace. After all, the site helped pioneer many of the exploitative practices I cover in this book. In its heyday, Myspace was the greatest expression of new media capitalism, and its sale to News Corp for more than $500 million in 2005 was seen by industry pundits at the time as a savvy purchase in light of the seemingly endless self-commodification of that site's users.

However, Myspace's failure and the concomitant rise of Facebook does provide a moment to nuance the critiques of social media explored in this book as well as do comparative analyses of both sites' softwares, architectures, and cultures. As scholars of science and technology studies (particularly in the social construction of technology school) argue, a critique of a failed technology allows us to avoid reifying successful ones as the natural result of technological development.[9] Hence, a critique of Myspace at this stage in its existence is a chance to avoid technological determinism. Furthermore, given the impermanence of websites, a postmortem of Myspace—at least the Myspace that existed before Justin Timberlake recently helped reinvent it—serves to somewhat preserve its memory for the public record.

Thus, this chapter contrasts the dominance of Facebook in social networking services (SNS) with the concomitant failure of Myspace. I argue that Myspace failed to associate its users, software, and third parties because it failed to produce an effective "real software abstraction." I first develop this idea by synthesizing the software engineering concept of abstraction and the Marxian concept of the real abstraction. Next, I compare Myspace and Facebook at the levels of aesthetics, code, culture, and appeal to marketers. I argue that instead of creating an architecture of abstraction in which users' affect and content were easily reduced to marketer-friendly datasets, Myspace allowed its users to create a cacophony of "pimped" profiles that consistently undermined efforts to monetize user-generated content. In contrast, Facebook has proven to be extremely efficient at reducing users to datasets and cybernetic commodities, all within a muted, bland interface that does not detract from marketing efforts. In other words, Facebook has associated social and technological elements into a real software abstraction, thus managing its users as immaterial laborers in the affective online marketplace.

Two Forms of Abstraction: Software Engineering and Marx's "Real Abstraction"

Software Abstraction

An important foundational document in software engineering is Frederick Brooks's classic tale of the development of the IBM System/ 360, *The Mythical Man-Month*.[10] Its publication in 1975 provided an influential explication of the nascent field of software engineering, drawn from the heart of what was the most powerful computer company in the world.

Brooks's argument about the production of software centers on a fundamental conceptual division between the *architecture* of the program and its *implementation*. "By the *architecture* of a system," he writes, "I mean the complete and detailed specification of the user interface. For a computer this is the programming manual. For a compiler it is the language manual. For a control program it is the manuals for the language of languages used to invoke its functions. For the entire system it is the union of the manuals the user must consult to do his entire job."[11]

The architecture is thus the surface of the program, the interface, the layer that the user will work with. As the spatial metaphor implies, "architecture" defines the space of the program: within it, certain things are possible. What is impossible is not to be contained in any shape in its architecture. For a software project to be successful, the architecture must be clearly specified early and act as the guiding document for all the tasks of realizing it.

Once the architecture is specified, implementation begins. Implementation involves what is popularly thought of as the work of making software: coding. Brooks argues that this is when software is "realized." Implementation is achieved by coding the interlocking components that will enable the architecture to function as specified. As such, whereas architecture is ideally rigidly specified, implementation is largely open-ended: one might use a variety of coding techniques or languages to achieve the functions described in the architecture. However, what is not allowed in Brooks's prescription is multiplication of functions and features; he uses the Cathedral of Reims as an example, because the centuries-old structure has a "unity" that "stands in glorious contrast" to other cathedrals that contain multiple architectural styles. Reims achieved its integrity "by the self-abnegation of entire generations of builders, each of whom sacrificed some of his ideas so that the whole might be of pure design."[12] Brooks argues that implementers need to similarly subsume their desires for individual expression to maintain the integrity of the architecture.

Contemporary software designers have built on Brooks's architecture/implementation hierarchy by refining the organization and planning of software systems. At the heart of contemporary thinking lies the concept of *software abstraction*. As Timothy Colburn and Gary Shute explain, software abstraction has its roots in Lockean philosophy. Like the distinction between "mountain" and Mount Everest, abstraction in computing means shifting from the particularities of the machine (the specific configuration of its hardware) to general software that works on that hardware:

> At a basic level, software prescribes the interacting of a certain part of computer memory, namely the program itself, and another part of memory, called the program data, through explicit instructions carried out by a processor. At a different level, software embodies algorithms that prescribe interactions among subroutines, which are cooperating pieces of programs. At a still different level, every software system is an interaction of computational processes. Today's extremely complex software is possible only through abstraction levels that leave machine-oriented concepts behind.[13]

The machine-orientated concepts that are left behind are the material, electronic events that always happen in modern computers: "Whatever the elements of computational processes that are described in textual programs . . . they are never the actual, micron-level electronic events of the executing program; textual programs are always, no matter what their level, abstractions of the electronic events that will ultimately occur."[14] Ultimately, computer programmers use abstraction to hide the material machine behind increasingly complex layers of code, which become a stack of abstractions, with concrete lower ones hidden by the more complex layers on top.[15]

This practice is now codified in the most important professionalizing document in software engineering: the *Guide to the Software Engineering Body of Knowledge* (commonly known as the *SWEBOK*), produced by the Institute of Electrical and Electronics Engineers (IEEE) Computer Society Professional Practices Committee. Per this guide, the overall architecture—the highest level of abstraction—is the blueprint for the entire project: "A software design (the result) must describe the software architecture—that is, how software is decomposed and organized into components—and the interfaces between those components. It must also describe the components at a level of detail that enable their construction."[16] This highest abstraction hides its internal, heterogeneous details, which are often expressed as modules (themselves conceived of as abstractions). The particular, line-by-line labor of coding is always secondary; it is the process of building modules within the larger architecture. Ideally, code should never get in the way of the architecture; it should only serve it.

However, the roots of abstraction lie in not only Lockean philosophy or the instrumental need to simplify complex engineering tasks but also the division of labor inherent in large software projects built in informational capitalism. As Barbara Liskov and John Guttag note, "The basic paradigm for tackling any large problem is clear—we must 'divide and rule.'" Their method to "divide and rule" is to plan the software project in a way that allows individual coders to work "independently with a minimum of contact," thus alleviating the inefficiencies of organizational communication. This is achieved by "decomposing" the architectural abstraction into modules, small programs that "interact with one another in simple, well-defined ways. If we achieve this goal, different people will be able to work on different modules independently, without needing much communication among themselves, and yet the modules work together."[17] For them, the act of constructing a large program and the act of managing labor are one and the same; decomposing abstractions into modules is the same as decomposing a large workforce into individual workers or small teams who answer to the architects of the system.[18] This division and hierarchy must be part of the overall organization of the company producing the software. As software engineer Phillippe Krutchen explains, "The architectural design approaches need to be supported by a matching organization that takes architecture as a key function, and understands its value and how it flows into other areas, such as planning, project management, product management, design or deployment."[19] For these software engineers, architecture is the infrastructure that determines how the software and the organization of labor function. Since software production remains a labor-intensive process (despite the development of compilers and higher-level languages), management of labor is crucial. As Brooks puts it, "Deliberate, even heroic, management actions are necessary to achieve coherence."[20]

Marxian Real Abstraction

While there appears to be a clear division in software abstraction between its loftier philosophical aspects and its down-to-earth management of labor, if we draw on Marxian critiques of abstraction, it becomes clear that these elements are mutually constitutive. Abstraction in software engineering is built on the foundations of abstraction in Western thought since the Enlightenment, and its roots lie in the rationalized management of labor. To explore this, I turn to Alfred Sohn-Rethel's *Intellectual and Manual Labor*, which provides a clear analysis of abstraction as both an ideal and material practice.[21]

Starting with Karl Marx's assertion that "it is not the consciousness of men that determines their being, but, on the contrary, their social being that determines their consciousness,"[22] Sohn-Rethel argues that ideal abstractions arise because of concrete actions in social formations: "Abstraction can be likened to the workshop of conceptual thought and its process must be a materialistic one if the assertion that consciousness is determined by social being is to hold true. A derivation of consciousness from social being presupposes a process of abstraction which is part of this being."[23] This leads to Sohn-Rethel declaring the money-commodity to be a "real abstraction," a contradictory phenomenon that "exists nowhere other than in the human mind but it does not spring from it. Rather it is purely social in character, arising in the spatio-temporal sphere of human interrelations. It is not people who originate these abstractions but their actions."[24] As commodity exchange gave rise to the universal equivalent of money, it also gave rise to abstract thinking. Since the money-commodity is timeless and socially disconnected to whatever material represents it, it allows its users to imagine other possible ideal formations that are timeless and diffuse as well as divorced from the rhythms of the natural world. The money-commodity thus becomes a second nature, as real as both primary nature and human action and just as capable of shaping human social relations. Here, in the overdetermination between the concrete and the abstract, we can see how something like the money-commodity can be a "real abstraction."

Sohn-Rethel's recent interlocutor Alberto Toscano has modified the former's theory to address cognitive capitalism. According to Toscano:

> Society is above all relation: the role of these univocal simple abstractions—such as value, labor, private property—in the formation of the concrete must be carefully gauged so that they do not mutate back into those powerless and separate, not to mention mystifying, intellectual abstractions that had occupied the earlier theory of ideology. But these abstractions are not mental categories that ideally precede the concrete totality; they are real abstractions that are truly caught up in the social whole, the social relation.[25]

Thus, Toscano draws on Sohn-Rethel, Louis Althusser, Slavoj Žižek, and Roberto Finelli to expand the concept of real abstraction to any ontological phenomenon by which "capital qua substance [becomes] 'Subject.'"[26] This includes familiar points of entry into historical materialist analysis: abstract labor, the commodity, and money. As with any conceptual expansion, debate over the primacy of any one real abstraction is possible. However, regardless of the primacy of one real abstraction (say, the money-commodity) over another (say, abstract labor power), the effects of any real abstraction include material consequences. In any case, real abstractions express themselves in social organization and are expressions of social organization. As Matthew Fuller and Andrew Goffey argue, "Abstractions, such as those produced by statistics, even those which are most often criticized for reifying or occluding a relationship to the real can, by such means, induce the emergence of a topological continuum and hence traverse scales of reality, introducing probabilistic, determining or contingent effects."[27] Such abstractions are real because they are actions; they are abstractions because they become part of the immaterial constitution of a whole way of life.

Thus, we see that software abstraction is not simply a method to mentally and ideally conceive of a software system; it functions as a real abstraction, something that is paradoxically both ideal and concrete, expressing itself in concrete effects. Software abstractions, like real abstractions, are an effect of capitalism, a system about which Marx argues, "Individuals are ruled by abstraction, whereas earlier they depended upon one another."[28] The architectural specifications of a software project are, in fact, capital qua substance becoming "Subject." This subject organizes labor to produce software commodities. Coders—intelligent, relatively autonomous human laborers—become objects producing object-oriented software systems for the benefit of their employers.[29] They are ruled by the software abstraction, the architecture. Just as Sohn-Rethel sees Immanuel Kant's "pure reason" as an artifact of mental/material labor split (and as an artifact of class division), so too is the architect/implementer division an artifact of class, with the same power to dominate workers as Frederick Taylor's division of conception and execution—even as it provides implementers with some autonomy to creatively write their code.

In the next section, I synthesize the software engineering concept of abstraction with the Marxian idea of real abstraction as these two concepts appear within social media and especially within the practices of user-generated content.

Real Software Abstractions in Social Media

Certainly, the progression from the nascent professionalizing document *The Mythical Man-Month* to the highly structured *SWEBOK* is not linear, nor has its attendant vision of labor in software production gone unchallenged. An obvious challenge to the proprietary mode of software production used by IBM, Microsoft, and Apple has been the open-source movement, led by such programmers as Richard Stallman and Linus Torvalds. Eric Raymond's "bazaar" model is perhaps the quintessential riposte to Brooks's architecture/implementation concept.[30] Raymond explicitly contrasts the authoritarian model of cathedrals and proprietary software with bazaars and open-source software. Whereas an architecture/implementer system hinges on the division of labor I have described—a division of labor Brooks sees as necessary to creating any coherent artifact—the bazaar model hinges on many people contributing freely to a project according to their abilities and in nonpredetermined amounts of time, just as small shopkeepers may set up tables and peddle their wares in a public space. In the most idealized form, the open-source "bazaar" model requires no central arbiter or authority; rather, every contribution is simply judged by one criterion: does it work? This model has been notably used in the production of GNU/Linux, which is a free, fully fledged, nonproprietary alternative to Windows or Mac OS.

The open-source model of production has greatly influenced Tim O'Reilly's concept of "Web 2.0," an organizing concept that has largely informed contemporary social media.[31] Web 2.0 arises, O'Reilly argues, out of the larger "architecture of participation" found in open-source software systems, such as GNU/Linux and the World Wide Web. These systems are built to be open to inspection and alteration. Their openness allows for anyone with the time and technical ability to contribute to them and shape them to heterogeneous needs. Web

2.0 is marked by user-generated blogs, wikis, and Web pages, all of which indicate a fundamental shift from closed, industrial modes of software and content production to a user-led, "produsage" model.[32] The "2.0" in Web 2.0 indicates that, on the surface, there appears to be a clean break between the old models of media and software production and the new ones marked by freedom, personalization, and participation. Once we update our software from version 1.0 to 2.0, there is no going back.

This appears to undermine my argument for the *continuity* of software production models and management of labor by capital by way of the real software abstraction. How do we move from industrial software development to the chaotic, user-led and -created world of Web 2.0 and social networks? How can anyone argue that the real software abstraction, which clearly functions to manage the labor of coders in firm-based software production, has been carried into social media, where the users are in control? After all, prior to the widespread use of computer networks, the *professional* (that is, not "user-led") practices of software engineering were limited to software firms, such as IBM and Microsoft.

The answer lies in the shift from software for personal computers to software built specifically for the Web. The advent of Web 2.0 and cloud computing in the 2000s introduced the "Web as Platform,"[33] shifting the emphasis of developers from building iterations of software for computer platforms to the "perpetual beta" model of development for the Web. A quick contrast between Microsoft Word and Google Docs illustrates this concept: while Microsoft builds Word as discrete versions to be released every few years (Word 95, 98, XP, 2007, and so on) and used on a personal computer, Google's software engineers release small changes to Docs on a scale of days, not years, and Google Docs works on nearly any Web-enabled device. A user might upgrade a computer to take advantage of the latest version of Word, but there is less need to think in terms of hardware when it comes to Web-based software, because Web-based software exists independently of any given personal computer or smartphone.[34]

However, the most salient change that has come with the shift in platforms lies in the ways in which the labor of software implementation is breaking out of the firm and being "crowdsourced"[35] on the Web. In an ideal social media site, the user is conceived of as an *implementer*, a laborer responsible for realizing the architecture conceived of by the site owners and designers. As the authors of the O'Reilly Media tech book *Web 2.0 Architectures* argue:

> A key trend in Web 2.0 is the inclusion of the user as a core part of any model. Most Web 2.0 examples have breached the purely technical realm and include users as an integral part of their workflow. Online applications are more than mere software; they represent a process of engagement with users. Users provide key functionality and content in most Web 2.0 applications, helping to build a web of participation and collaboration.[36]

The "engagement" of the user could be thought of in terms of "interactivity" or "participation," as Andrejevic[37] has so aptly described. But it also could be thought of as implementation. Web 2.0 sites are essentially empty frames: imagine Twitter, Facebook, or YouTube without any user-generated content. But rather than dwell on what's missing, the frame itself should be examined. It is, in fact, an *architecture*, waiting for a user to realize it with content.

The ideal social media site is thus a real abstraction, a software abstraction capable of directing users' concrete labor, social relations, and interactions. Each user labors over a small portion of the site, exchanging bits of personal data for access to the service and to other users. *Web 2.0 Architectures*, like Brooks's classic text and the *SWEBOK* before it, offers an idealized vision of software that has real effects on the organization of labor needed to implement it—the only difference between social media's underlying software and prior iterations of software is the role of users as implementers (in addition, of course, to the waged laborers who work for these firms and create the frames themselves by coding them in Javascript, HTML, PHP, and so on). Thus, social media architectures are *real software abstractions*.

This concept of real software abstractions informs the distinction between Myspace and Facebook and explains—at least in part—how the latter has thrived and the former has become a latter-day Pets.com.

Myspace's Concrete Chaos versus Facebook's Abstractness

For critics of social media and free labor, Myspace has functioned as an exemplar of exploitation and panoptic surveillance.[38] As Mark Coté and Jennifer Pybus argue, Myspace is a biopolitical disciplinary system meant to train its users to be immaterial laborers: "The 'work' of Myspace, as a corporate entity, is to 'monetize' [the affective work of users] in a manner which does not compromise the good will of users."[39] However, Myspace's market failure and Facebook's rise require us to nuance these critiques. In spite of Myspace's power to create and discipline its users as immaterial laborers and sell their data to marketers, why did Myspace ultimately fail?

A simple aesthetic comparison of the sites reveals a path to an answer. Consider a subculture that may not have too much obvious appeal to marketers and advertisers: Satanists. Facebook's openness allows for users to express a wide range of interests, including Satanism. The Satanism Facebook page features an inverted pentagram as its profile image.[40] The pentagram is white on a black field; it is a clean SVG file, with no pixilation. As such, it fits into the overall aesthetics of Facebook: sans-serif fonts, clean blue banners, rounded edges, black text on a white background. Next to the pentagram is a "Like" button, a binary switch that allows Facebook users to signal their affiliation with Satanism. As of this writing, more than forty-nine thousand Facebook users "like" Satanism. Below this "Like" button is an excerpt from Wikipedia, pulled in with Javascript and neatly displayed in its own HTML div in the center of the profile. According to the Wikipedia excerpt, "Satanism is a religion that is composed of a diverse number of ideological and philosophical beliefs and social phenomena. Their shared features include symbolic association with, admiration for the character of, and even veneration for Satan or similar rebellious, promethean, and liberating figures." The Wikipedia description, arising as it has out of that encyclopedia's "Neutral Point of View" policy, is clinical and detached, fitting well with the clean interface of Facebook. In terms of code, the use of Javascript to pull in external content, such as the Wikipedia excerpt, along with highly standardized HTML and CSS, results in sixty-eight lines of code; again, even at the code level, Facebook is well organized. In all, the Facebook Satanist page, like all Facebook pages, is so aesthetically clean that it renders the content in the frames nonthreatening; one imagines the most subversive content being muted within this architecture, reduced to a series of HTML divs, smooth images, clinical text, and well-structured code.

On the other hand, before its recent redesign, Myspace's users had constructed a wild, cacophonous array of "pimped" pages, using coding hacks to radically alter the layout and settings of their pages. Staying with the Satanism theme, we examine Satan's own Myspace page as it appeared before Myspace was sold off by News Corp.[41] While Myspace's original default layout was similar to Facebook's in that it used blues and whites, Satan's page was red text on a black background. The profile picture featured Satan using Jesus Christ as a slingshot; he was aiming directly at the viewer.[42] Although Myspace did have standardized layout and navigation (even for many "pimped" pages), in my browser, the Myspace navigation bar (an element that in practice would have been on every page to standardize navigation throughout the site) was gone, replaced with a gray space at the bottom of the page, indicating a bug in the customized code written to override that standardized element. The various HTML divs on Satan's page varied in width; thus they did not line up neatly along the left side of the screen. Likewise, Satan's friends' avatars came in multiple sizes, and thus the collection of avatar images was uneven. While the Facebook page offers the viewer the option to "Like" Satanism or not, even though I was not logged into Myspace and I had never "friended" Satan, I saw that "Satan is in your extended network." In Myspace, "like" it or not, I had no choice but to be connected to Satan. As Satan informed me, this member of my extended network was dedicated to "indulgence instead of abstinence" and "kindness to those who deserve it instead of love wasted on ingrates!" This language was a far cry from the clinical discussion of Satanism found in the Wikipedia excerpt on the Facebook page. Finally, at 335 lines of HTML, CSS, and JavaScript, this pimped Myspace page used nearly five times the code of the Facebook page, rendering it slower in my browser, especially since the code was interfering with the Myspace navigation bar. In sum, during its heyday between 2005 and 2011, the architecture of Myspace was constantly restructured by users who relied on the pimped codes to radically alter their pages, and, as danah boyd has argued, these users often expressed "dangerous" ideas, such as black and Hispanic racial identities and raw, working-class sexuality, in their profiles.[43] This was a far cry from the aesthetically muted Satanist Facebook page.

Ultimately, however, aesthetics is not the key to the questions raised by Myspace's failure, but aesthetic analysis points toward further areas to examine. The aesthetics of Myspace arose from the users' ability to inject customized code to alter the default settings. Beyond simply altering the layout of their profiles, Myspace hackers were also able to write code that hid friend lists, the "Last login" field, recent status updates, and, most importantly, advertisements.[44] Although the last alteration was a breach of the Myspace Terms of Service, the sheer number of sites providing the code to do so pointed to a basic antagonism within the site (and broadly on the Internet in general): many users resist advertising in any way possible. In Myspace, coding hacks made it possible; Facebook offers no such workaround.[45] Although all for-profit social networking sites strictly prohibit users from hiding advertisements on their profiles, the very possibility of ad blocking makes marketers extremely nervous.[46] Without advertisements, their logic goes, these "free" services cannot possibly exist.

In addition, consider the simple fact that Satan has a page, even though the actual existence of Satan is highly doubtful. This points to another architectural failure of Myspace: the rampant use of false profiles. Like Satan, a wide range of fictional characters had profiles on Myspace. However, beyond fictional characters, users often have fake profiles for a variety of purposes. In this way, Myspace duplicated the practices—and many of the problems—of Friendster, an SNS that preceded Myspace and has since failed in the market. Users of Friendster often created fake accounts, "Fakesters," to challenge the boundaries of the Friendster architecture.[47] The "Fakester" phenomenon lead to problems for Friendster, since that site never intended to allow fake profiles and worked vigorously to delete them. Similarly, the ease with which a user could create a fake Myspace profile resulted in widespread moral panics about sex predators, defamation, and "cyberbullies." For example, a sex-predator panic centered not on fake profiles but on *Wired* magazine writer Kevin Poulsen's correlation of data on registered sex offenders and Myspace profiles. Poulsen found 744 registered sex offenders on Myspace but speculated that more could easily be using fake profiles on the site to avoid detection.[48] Spurred by this revelation, pundits began to argue that adults were simply lying about their ages and other personal details to gain access to children on the site.[49] Similarly, several lawsuits were brought against users who created parody profiles of real people to mock them.[50] Finally, the suicide of Megan Meier, a result of her being harassed on Myspace by another user with a fake profile, led to proposed legislation against "cyberbullying."[51]

These incidents resulted in press coverage that heavily relied on technological determinism to argue that Myspace was the cause of increased sex predation, defamation, and bullying. While this determinist argument is clearly faulty—these phenomena predate Myspace—this coverage added to that site's image as a dangerous space.[52] These highly covered incidents tended to overshadow another prevalent subset of fake profiles: spam profiles, often carrying either advertisements (usually for porn sites) or malware, such as trojans or viruses.

In contrast, while Facebook certainly has its share of fake user profiles,[53] its history as a closed network has led to its current user culture of preferring real-world identities to fantasy identities.[54] Rather than being an open service like Friendster and Myspace, Facebook grew slowly by adding more colleges, then workplaces, then high schools, until finally opening up to the general public in 2006. Each of the stages prior to opening up to the public relied on third parties (namely schools, initially Ivy League ones) to vet the identities of its users. In essence, real-world identity was built into Facebook's architecture. The link between real-world and online identities has proven extremely attractive to advertisers.[55] With Facebook, marketers can reach highly specified demographics, customizing advertisements along myriad, abstract categories: age, gender, sexual orientation, location, work history, and purchase history, with all these data points linked back to users' real-world identities. This move has been so successful that Google is (as of this writing) attempting to replicate it by linking YouTube comments (which had been pseudonymous) to Google+ profiles, which are more commonly tied to users' real-world identities.

In addition, Facebook's Connect service is built on this heritage of real-world identities; such sites as CNN, Disqus, and the *New York Times* use Facebook Connect as their users' de facto online identity cards. Facebook Connect's role in vetting online identity is so effective that pundits argue that it is superseding open-source identity efforts, such as OpenID, as well as potential government-backed ID systems, such as that proposed in the *National Strategy for Trusted Identities in Cyberspace*.[56] Thus, Facebook's massive database of real-world identities is connecting well to marketers who desire increasingly granular data on potential, identified customers. Even as Facebook users click away from Facebook.com, they can remain logged in to the SNS, and thus their visits to other websites can be tracked, with the resulting data stored for later analysis. Facebook's public success in creating a Web-based identity protocol further increased its overall dominance over its competitor Myspace.

These factors reduced advertiser interest in Myspace, even as Rupert Murdoch's News Corp was ramping up advertising sales efforts. In a profile in *Advertising Age*, Gavin O'Malley[57] explains marketers' reluctance to advertise in the "less structured" areas of the site—that is, user profiles—opting instead to concentrate on "structured" areas, such as Myspace Videos and Myspace Music. "Less structured" is, of course, a euphemism for user-created profiles, popularly perceived to be a dangerous space. Returning to my Satanism theme, a user-generated McDonald's Myspace page featured a user-generated image of Ronald McDonald with "Satan" emblazoned across his chest in the comments area, hardly the image that the corporation wants to cultivate.[58] Moreover, as the site's traffic dwindled, advertisers increasingly began to consider it a "relic," especially since advertising executives themselves were abandoning it.[59] A vicious cycle continued to drive Myspace traffic—and thus revenue—further down. In 2010, Myspace earned $347 million in advertising revenue; in 2012, it lost $43 million.[60] In comparison, Facebook exceeded $1 billion in 2010,[61] doubling that by the end of 2013.[62]

Myspace's Abstraction Failure

Thus, I see in the rise and fall of Myspace what software engineer Joel Spolsky calls "abstraction failure,"[63] a lawlike reality of computing. In the technical sense that Spolsky is concerned with, this simply means that high-level abstractions, such as the operating system graphical interface, will eventually malfunction: icons might not behave as they are expected to, screens could go blank without warning or recourse, or (to use an example quite common in Windows 98 and XP in the late 1990s and early 2000s), the "Blue Screen of Death" might appear, warning the user that "A fatal error has occurred. To continue, press Enter to return to Windows or Press CTRL+ALT+DELETE." In these cases, users are confronted with deeper, vestigial, and, most importantly, concrete layers of the computer: faulty memory chips, failing hard drives, poorly written driver files, endless and unbroken loops in software. Moreover, the problem is always at the level of the physical machine: electrons are out of their intended places. To put it another way, the machine (and hence the material) bubbles up through the layers of abstraction and confronts the user.[64] Concrete, specific problems interrupt the smooth layering of abstractions on top of the machine.

But Myspace's was not merely a technical failure; it was a real abstraction failure. Beyond the annoying technical glitches of software abstraction failure, real abstraction failure involves a failure to discipline labor effectively, a failure to direct laborers' concrete actions, and thus a failure to capture the surplus value of that labor. At the interface level, Myspace's architects violated the professional practices of software engineers: control your architecture and allow the implementers to realize it, but never allow implementers to add features willy-nilly. Indeed, Myspace failed to discipline its users into producing content that adhered to the designs of the site owners. Myspace users' rampant employment of CSS and HTML hacks, embedded video and audio files that run as soon as a user's profile loads in a browser, and the increasing amounts of spyware on the site led *PC World* to declare it the worst website of 2006—even at the height of its popularity.[65] The false profiles and subsequent moral panic also contributed to the site's dangerous aura. Although Web 2.0 has been heralded as more emancipatory and democratic than traditional mass media, the lesson of Myspace is that too much emancipation frightens advertisers. The faults of Myspace—when viewed from the perspective of new media capitalism—is that it was too particular, too concrete, too heterogeneous to be contained. In sum, the concrete, chaotic, freewheeling desires of users bubbled up through the architectural abstraction to confront Myspace's intended market of advertisers, and the advertisers did not like it, resulting in a "fatal error."

Facebook, on the other hand, has done a remarkable job of disciplining its users. Its rigid layout and its clean architecture are artifacts of its intent to and continued success in directing its users via abstraction, and, as a consequence, abstracting value from the aggregated labor of its users. Users' affective labor—their "likes" and "social graphs"—become reduced to commensurable datasets, reflecting what Eva Illouz calls "emotional capitalism": "Never has the private self been so publicly performed and harnessed to the discourses and values of the economic and political spheres."[66] This reduction of inner, subjective life to the cold logic of exchange is built into the system of capitalism. Returning to Sohn-Rethel, as subjective as use-values are (and what is more subjective than emotion?), they must be reduced to commensurable exchange-values in the process of exchange; otherwise, exchange fails: "The actions of exchange are reduced to strict uniformity, eliminating the differences of people, commodities, locality, and date."[67] Everything is reduced to the universal equivalent: "This uniformity finds expression in the monetary function of one of the commodities acting as the common denominator to all the others."[68] Whereas Facebook is capable of monetizing emotion, Myspace, despite its best efforts, failed to make this reduction.

Of course, Facebook's ability to attract far more users than Myspace is surely a key to its success. Returning to the work of boyd, a major cause of this success may very well be the moral panics surrounding Myspace: sex predators, raw sexuality, and "dangerous" expressions of racial identity. However, expand the architecture/implementer metaphor and another cause is revealed: Facebook was simply easier for users to implement. Whereas Myspace's users' pimped pages created inconsistent design and navigation, making it difficult to load a user's page, find basic information about that user, and then make a social connection, Facebook's standardized system is far simpler: click "like," click here to add a friend, click here to add a picture. We might compare the playwork of implementation in Facebook to the ludic and yet highly constricted possibilities of identity formation made possible by consuming in a clean and bright suburban shopping mall versus the anxiety felt (at least by elite whites) in shopping in malls on the "bad side" of town. In a sense, the implications of boyd's work (i.e., that educated whites fled Myspace for Facebook) meets my own in that mass audiences have chosen a social network that does not inadvertently present them with the Other or ask them to do complicated work. Facebook reduces these concrete complexities into a clean, white template that awaits user implementation.

Conclusion: Facebook's Potential Abstraction Failure

However, Facebook is not necessarily a flawless machine. Facebook's overhyped and ultimately stumbling IPO reveals yet another abstraction failure. As Sohn-Rethel notes:

> People become aware of the exchange abstraction only when they come face to face with the result which their own actions have engendered "behind their backs" as Marx says. In money the exchange abstraction achieves concentrated representation, but a mere functional one—embodied in a coin. It is not recognizable in its true identity as an abstract form, but disguised as a thing one carries about in one's pocket, hands out to others, or receives from them.[69]

For a brief moment, the news coverage of the vast volume of investment flowing into Facebook revealed the exchange abstraction, resulting from the aggregated actions of nearly a billion users, actions performed "behind their backs" or, more properly, behind the glossy interface of Facebook. While Marx cautions us not to confuse the money-commodity with the abstract process of exchange—that is, do not confuse gold with the abstraction it represents—for the millions of Facebook users, there should be no confusion, since the wealth flowing through Wall Street banks is abstract enough to most of them. It is a "concentrated representation" peculiar to our historical moment: a speculative bubble, built out of the commodification of the private lives of Facebook members. Thus the coverage of the amount of the money flowing through the site and into Wall Street financial firms is another form of abstraction failure, one that starkly and undeniably reveals the power of Facebook to appropriate the emotional lives of its users, even as investors are betting that Facebook can hide this fact from its users behind an ideology of participation and connection.

In the final analysis, it is very important to remember that real abstractions require such ideologies to support them, and thus *real abstractions can be dissociated by critique*. Just as the money-commodity is ultimately meaningless without social faith in it, our social connections need not be mediated by a site like Facebook. Users may have flocked to it because of its promise of self-expression and social connection. They may have abandoned Myspace because of its concrete chaos and "dangerous" race and working-class expressions. And, for a brief moment in history, Facebook may succeed in convincing users that it is simply providing a service and not acting as an advertising spy network. But that top layer of software abstraction can be torn aside, revealing the mechanics of exploitation hidden just beneath. As Alfred Whitehead argues, "You cannot think without abstractions; accordingly, it is of the utmost importance to be vigilant in critically revising your *modes* of abstraction. It is here that philosophy finds its niche as essential to the healthy progress of society. It is the critic of abstractions."[70]

Returning to the real software abstraction and the software engineering idea of "abstraction failure," we can see that critique of abstraction is not limited to philosophers and academics. It can be the province of users who recognize the hidden purposes of the software they work with. As Oliver Leistert asks, "How much direct marketing are

Facebook users willing to take? How many drastic top-down changes of the user's Facebook experience are possible unless they understand that their presence on this site and what they do there is in tension with the company's goals that provides this digital environment?"[71] These questions orient us to changes in the real software abstraction that will alert Facebook users of the ways in which their emotional lives are being commoditized. The bungled IPO, Zuckerberg's frighteningly messianic language about making absolutely everything open and having all information flow through Facebook,[72] and the privacy train wrecks[73] have led to users critiquing Facebook not in terms of aesthetics but in terms of exploitation and profit.[74] Thus, once again, software engineering presents us with a window into the processes that social media comprise and thus a means to critique them and, as I explore in the final chapter, reverse engineer them.

4

Standardizing Social Media

Technical Standards, the Interactive Advertising
Bureau, and the Rise of Social Media Templates

> It is claimed that standards were based in the first place
> on consumers' needs, and for that reason were accepted
> with so little resistance. The result is the circle of
> manipulation and retroactive need in which the unity of
> the system grows ever stronger.
>
> —Theodor Adorno and Max Horkheimer, "The Culture
> Industry: Enlightenment as Mass Deception"

> The best minds of my generation are thinking about
> how to make people click ads. That sucks.
>
> —Jeff Hammerbacher, quoted in Ashlee Vance,
> "This Tech Bubble Is Different"

When it comes to discussions of the history and politics of social media, technical standards are, oddly enough, downright sexy.[1] Transmission Control Protocol/Internet Protocol (TCP/IP), the communications standards that structure the Internet, have been pointed to as the source of the Internet's politics of academic freedom and entrepreneurialism.[2] Hypertext Markup Language (HTML) and Hypertext Transfer Protocol (HTTP), the standards that structure the Web, are touted as the source of the Web's meteoric growth and thus its fundamental challenge to mass media.[3] BitTorrent and other decentralized peer-to-peer technical standards are proclaimed to be one of the most valuable tools for undermining central authority and enabling innovation and user-led production.[4]

Many scholars and pundits have consistently argued that these protocols are major determinants of the politics made possible by social media. The argument: since the communications protocols underlying the Web are distributed and decentralized, social media built on top of the Web are almost necessarily distributed, decentralized, and thus democratic to their core. Along with concurrent technological developments, particularly the "bazaar" style of open-source software production,[5] social media are presented as a democratizing force. In some extreme views, political protests and revolutions are branded "Twitter Revolutions" and "Facebook Revolutions," made possible (only) by SNS standards, "Web-as-platform"-based Application Programming Interfaces (APIs), and the network effects of those social media.[6]

This chapter takes its inspiration from studies of social media that connect technical standards and technological politics. However, I take a different tack: I discuss the less-acknowledged role of *advertising* technical standards in the history of social media. Part of what enable the movement of data from one site to another are, of course, communications protocols and standards. Advertising standards form part of this equation and should be seen as another (metaphorical) layer in the stack of network protocols. Certainly, advertising standards are just as distributed and decentralized as, say, TCP/IP or HTTP. They are produced by consensus—theoretically, anyone can contribute to their design—just like any protocols established by the World Wide Web Consortium (W3C). And the particular advertising standards consortium I examine here, the Interactive Advertising Bureau (IAB), is remarkably open about how its standards work and what they are intended to do, just like any public goods–producing standards consortium should be.

However, in our euphoria over networks, decentralization, participation, and consensus, we often forget that standards are by no means inherently democratic. At the very least, advertising standards produced by the IAB have in part determined the rise of what Kristin Arola calls "template-driven"—that is, standardized—social media, such as Facebook and Twitter.[7] As heterogeneous as the *content* of these services is, their *forms* (or maybe I should draw from protocol-speak and say "wrappers") are rigid and unalterable by users. Why? At first glance, we might say that this is simply because these sites are "user-friendly" and that the best way to allow many users to participate in the media is to produce push-button templates so that they do not need to worry about learning to code. As we saw in Chapter 3, Myspace died under the weight of its pimped, hacked, user-frightening profiles. But here I argue that these standardized templates are, in fact, glistening façades hiding what increasing numbers of media scholars recognize as the infernal machinery of surveillance. Moreover, I argue that social media surveillance itself is not possible without standards, because to be effective it relies on gathering standardized data and storing such data in rationalized archives. The IAB has produced the standards necessary for effective social media surveillance. Recognizing this, social media sites, such as Facebook, Google+, and Twitter—which live, of course, off capturing user data and selling it to marketers—deploy templates that adhere to IAB standards of data surveillance and prevent users from altering the mechanics of that process.

To demonstrate this, I start by exploring the role of contemporary standards consortia, outlining their typical three-part ideology. First, they present themselves as solving a *user problem*—that is, their production of standards is intended for a particular set of users and is aimed at producing a public good that alleviates the confusion of incompatible technologies, rapid obsolescence, and high research and development costs. Second, standards consortia argue that their work *creates new markets*. Once the user problem is solved, consumers feel more confident about investing in technology, and thus multiple companies can compete to win their favor with different designs and options. Finally, standards consortia draw from the larger discourse of neoliberalism to present themselves as privatized regulatory bodies that are far more responsive to technological change than are government regulators. This ideology of *self-regulation* is deployed both as a means to counter state-based regulation and as an invitation to firms to join the consortia and help set the rules by which they do business.

Next, I use this three-part ideology to examine the IAB's production of advertising standards. I argue that the IAB formed to solve a classic marketing *user problem*: the inability to gauge the return on advertising investments. The IAB's production of standardized advertising, ad-exchange networks, and surveillance technologies and metrics gave rise to a *new market*: contemporary social media. Social media's template-driven shape reflects the underlying standards produced by the IAB. Finally, I address the question of structure and agency by using the IAB's *self-regulatory* ideology as a means to discover possible avenues of resistance to this political, economic, and social structure.

Standards Consortia

Andrew Russell aptly describes standards as the "social process by which we come to take things for granted. Through standardization, inventions become commonplace, novelties become mundane, and the local becomes universal. It is, in short, the historical process by which discoveries are rendered into the material and immaterial substance of our everyday lives."[8] Standards are thus technical artifacts that are to be forgotten, ignored, or placed in the background, leaving us more capable of concentrating on more pressing concerns. For example, I can ignore the intricacies of the Internet Mail Access Protocol (IMAP) standard, because my e-mail client handles that for me. I do not think about the ways the tiny pits on a DVD are read by a laser in a player and converted into images and sounds. I do not worry about how electrical currents work when I plug in a reading lamp. In these cases, I forget the standard to do something else, something more creative or relaxing, such as write an e-mail, watch a movie, or read. This is the power of standards. The heterogeneous engineers of technology recognize this power, and thus they spend a lot of time producing standards so that the rest of us can forget about them and instead enjoy such pastimes as writing e-mail and watching DVDs.

Standards organizations thus are central to technological production and diffusion. In general, standards organizations "mediate a standards ideology"[9] by shaping "how standardization should proceed, beliefs on what is important in the process of establishing standards and why it is important, [and] assumptions about the standards environment."[10] Many standards organizations are government-sanctioned bodies, such as the National Institute of Standards and Technology, the American National Standards Institute, or the International Standards Organization, and promote an ideology of democratic participation and consensus.[11] These government-sanctioned bodies are artifacts of modernity, the larger projects of nations imposing order on the world via railroad gauges, time zones, telegraph languages, periodic tables, and engineering protocols.[12]

However, standards organizations have not been immune to the neoliberal push for privatization.[13] During the 1970s and 1980s, many private, industry-based standards consortia formed to create standards as a means of self-regulation and market management.[14] This is especially true in the areas of networked computing.[15] In fact, the Internet looms large in the mythology of standards consortia, because Internet standards are largely produced by nongovernmental organizations using open standards of interoperability and consensus-style decision making.[16] The production of Internet and telecommunications standards by private consortia is taken as evidence of the efficiency and nimbleness of private consortia, especially in comparison to the putatively slower government-sanctioned standards organizations.[17]

Private standards consortia are far less concerned with their responsibility to a government or adherence to ideals of democracy and consensus than are their state-sanctioned counterparts.[18] Rather, they hold themselves accountable to only their dues-paying members. While their intermediate goal, the production of technical or practical standards, is the same as that of state-sanctioned groups, the long-term goal of a private standards consortium is, in fact, a coordinated marketplace; the standard must be implemented by the majority of the business community. Otherwise, the work going into its production is wasted, and there is even the risk of a single company building a monopoly and thus authoring a de facto, competing standard. Thus, standards consortia are in large part devoted to the practice of building a network of like-minded, coordinated businesses as much as they are to the production of a standard.[19] The failure rate of this practice is high;[20] corporations that join standards consortia often abandon them as market opportunities arise. Moreover, standards consortia are often formed by corporations that seek to compete with other corporations' standards consortia.[21]

These two goals—the production of a standard and its wide implementation—can create tension in a standards consortium. The desire to get a standard developed quickly and the need to get it implemented widely can conflict. The first benefits from small gatherings of like-minded companies; the latter typically requires a wide membership base who might not contribute to the standard but will adopt it.[22] A successful consortium balances these competing interests with a mixture of democratic and oligarchical structures spread across local and global offices. It is capable of associating a network of technologists, business leaders, and members of nonprofits into a coherent organization.

To produce and maintain such a network, standards consortia often deploy three ideological frameworks for their work: they address a "user problem," they create a stable market for technologies, and they create regimes of self-regulation. In terms of the user problem, standards consortia present themselves as benevolent gatherings of corporations dedicated to solving typical technology headaches: lack of interoperability, poor technical support, and speedy obsolescence.[23] "User" might mean individual consumers,[24] but in many cases the term refers to industrial firms that purchase technologies.

Second, standards consortia present their work as producing a marketplace where all members can prosper by working together in some areas (i.e., in terms of technical specifications and compatible interoperation) while continuing to compete in others (i.e., for customers). As critical studies of consortia note, this approach is largely an effort to maintain stability in technology markets. Rather than allowing one company to gain a monopoly through a superior or widely adopted technology and thus create a de facto standard (as Microsoft did with MS-DOS and Office), companies band together to manage the production of standards and maintain their market positions.[25]

Finally, symptomatic of the neoliberal economic context in which they operate, standards consortia present themselves to governments as self-regulating and thus not in need of state-sanctioned regulation. Although the gathering of many corporations in a network of coordination might raise the specter of collusion and monopoly (at least, to any state that cares to regulate such a thing anymore),[26] standards consortia argue that a standard is a public good, available to anyone who wants to use it regardless of whether that individual contributed. Their work is thus explicitly political-economic: they present themselves as consortia of corporate citizens working for the betterment of all by simultaneously producing a public good *and* a marketplace where value can democratically flourish via creative competition.

The "User Problem" and the Rise of the IAB

In light of the attention that sexier standards, such as TCP/IP, have gotten, it is easy to overlook the role that advertising standards consortia play in social media. However, even a cursory glance at the member list of the IAB reveals how important advertising standards are. The IAB counts among its members big firms, such as Amazon, the *New York Times*, AOL, Google, and Facebook, as well as small marketing firms, such as Pulse Marketing, College Degree Helper, and WebFloss.

Why have these large and small companies come together? A place to start is the old advertising saw attributed to John Wanamaker: "I know half of my advertising [budget] is wasted; I just don't know which half." Borrowing an often used (and abused) term, computer networks are "disrupting" the traditional practices of advertising. After nearly a century of standardizing the practices of advertising in such media as print, radio, and television, and especially after shaping the state's regulations of marketing in these media,[27] the advent of the Web created new challenges to the marketing industry. While the practice of measuring the impact of mass media advertising had been developed for decades, the Web's architecture rendered many of those practices useless.[28] The central mythological crisis of contemporary advertising is that the user is now in control. Whereas in mass media, advertisers simply transmitted their messages ad nauseam to relatively passive audiences, new media allow for active audiences who are producers as well as consumers.[29] The Web in particular allows users to create their own media, control the flow of information they encounter, and, most devastatingly, avoid the interstitial advertising of mass media altogether.

And yet, early empirical examinations of the Web have found that users have often felt *out of control* in the face of a network of universal machines.[30] Websites varied wildly; even with the standards-setting body the W3C, the "browser wars" between Microsoft's Internet Explorer and Netscape's Navigator, coupled with the inevitable growing pains of any new medium, meant that users confronted a sometimes bizarre mediascape of sites "under construction," dead links, and pop-up ads. A common metaphor of the 1990s was that the World Wide Web was the "Wild Wild West."[31] In this space, interaction was as open-ended as many other human activities; uncertainty, surprise, and anxiety were the order of the day. Moreover, this wildness was a *product* of user agency: users were making sites as well as visiting them.

It is easy to forget that marketers and advertisers confronted the same chaos that everyday users did. For example, in the early 1990s, banner ads came in more than 250 different sizes, making it extremely difficult for advertising agencies to design ads for more than a few sites.[32] While Web metrics promised to reveal consumer habits in greater detail than any prior media system, there was no agreement on which metrics to use and how.[33] Even when online media companies had basic data on their visitors, they were hesitant to share the information with advertisers, causing distrust on both sides.[34] This antagonism can be explained in part by the overall uncertainty of how to fundamentally approach Web advertising: is it like TV? Print? A mix of media?[35] In short, the situation was worse than the old advertising saw: Web advertisers could not be sure that *any* of their advertising was effective. And yet, the hype over the potential for Web advertising was reaching a fever pitch as more and more millions of dollars—very soon to be billions—were being spent on online ads.

Thus, the situation—which was clearly a "user problem" of compatibility—cried out for standards that could reduce wasted marketing budgets and coordinate a new advertising industry. In 1996, representatives from CNET, Infoseek, Juno, Microsoft, Prodigy, Softbank Interactive, Starwave, Time Inc., and Turner Interactive formed the IAB[36] to solve the problems arising in new media advertising. The organization moved quickly. Within months, the IAB produced its first standard: banner-ad sizing, culling the number of sizes from more than 250 to 8.[37] These standards are still in use today (although banner ads are largely seen as ineffective). Through committees comprising representatives of member corporations, the IAB continued to standardize the size, shape, and behavior of advertisements, including pop-ups and pop-unders,[38] in-line units,[39] and so-called rich-media advertisements.[40] More recently, the IAB has sponsored contests for new, dynamic forms of Web advertising, accepting advertising sizes "Billboard," "Filmstrip," "Portrait," "Slider," and "Sidekick" as new standards.[41]

The standardization of ad sizes was a boon to the industry. Instead of having to redesign an ad to fit each and every website they wanted to display it on, marketers could create a single advertisement that could be displayed across multiple sites. For site designers, standardized sizes eased the creation of Web pages.[42] The standardization of sizes also allowed for a clear distinction between editorial and advertising content, allowing magazine and newspaper practices to be remediated on the Web. Size standardization thus was a key step in reducing the chaotic appearance of the Web. Along with advertisement standards, Web design books began promoting standardized structures for the millions of students who would take Web Design 101 in colleges around the world: three columns, navigation to the left, main content in center, the Golden Ratio, banner of so many pixels, use of div tags and CSS to divide and style content, and so on.[43]

More importantly, standardized sizes allowed for increased flexibility in the placement of ads, allowing marketers to adopt eye-tracking technology to the Web. Eye tracking allowed them to discover how to attract the most attention to advertisements.[44] It revealed that Web users simply did not look at the standardized banner ads at the tops of pages, a revelation that led marketers to experiment with different ad locations and formats.[45] This process of weeding out clutter and placing ads in high-attention sectors of websites had the consequence of reducing the supply of online advertising space. That is, rather than stick ads into any spare spot on the screen, Web designers who sought advertising revenue were able to use only a few areas of their pages.[46] Advertising became a "seller's market," at least on the most-trafficked websites (such as Yahoo! and the *New York Times*), and any company seeking to purchase advertising space had to become far more focused on producing return on investment.

Thus, contemporary Web advertising and design standards bring about what Michel Foucault might call an "instrumental coding of the body," a linkage between human gestures and movements and the material object they work on.[47] Although the Web is a site of subtler movements than the movements of the soldier and the gun Foucault describes, the principles are similar: the eye becomes trained to move through a certain path on a site, largely in a Z shape; the hand moves the mouse to navigation boxes and links; we learn where to look for content and where to look (or not) for advertisements. With enough time, such human-computer interactive codings earn the label "intuitive," a curious name implying that the actions are somehow innate when in reality they arise after hours of subtle training in the use of Web browsers and site navigation. Not only layouts but also the mouse and eye movements of users have become standardized.

The IAB's success with standardized advertising sizes, the subsequent rise of eye-tracking experiments in ad placement, and the decrease in Web space supply all point to the need for the most important standards the IAB created: precise and clearly defined metrics for data collection and analysis—the heart of the online advertising industry. The IAB's goal was not just the standardization of sizes and layouts; its members wanted to structure the very language that new media advertisers used as well as their methods for measuring success, thus providing a means for marketing to break out of individual websites and become an intersite practice.[48] The IAB began the process by defining Internet advertising terms, ultimately producing a glossary.[49] "Containing technical, measurement and business terms," the glossary "was amassed with the assistance of a broad base of expert advisors from advertising agencies, technology providers, publishers and research companies, and will be periodically updated by the group."[50] This process of standardizing advertising jargon simplified negotiations between publishers and marketers, and it led to a standardized legal advertising purchase contract now in use by such sites as *USA Today*, *TIME*, *People*, Disney, Shop.com, and AOL.[51]

But, of course, the IAB's data standardization could not stop with definitions of terms. The most heated debates in online advertising center on how precisely to measure advertising effectiveness and thus how to put a price on the audience's eyes, minds, and hearts. Whichever organization could create, implement, and control standard metrics for Web advertising would have much influence over the industry, and indeed many firms competed to create such a standard.[52] Rather than simply produce a standard in competition with other organizations, the IAB partnered with the American Association of Advertising Agencies (AAAA), the Association of National Advertisers (ANA), and PricewaterhouseCoopers (PWC) to produce standard metrics at the beginning of the 2000s.

PWC began the standards-production process by studying companies from three sectors of the online advertising industry: ad networks and servers (including DoubleClick), portals (including Yahoo! and MSN), and "destination sites" (including the *New York Times*, Forbes, and Disney).[53] Together, these three sectors form the basic architecture of advertising online, so PWC sought commonalities and discrepancies in terms, practices, and metrics among them: "A fundamental premise of this report is that in order to achieve reliable, accurate and comparable ad campaign measurement reporting, there must exist a set of standardized metric definitions that are applied to a well-controlled process."[54] In other words, a basic scientific method must be coupled with a standardized language and used across all the sectors of the industry. Although it identified many discrepancies, PWC found that all the companies studied used the same five metrics: ad impressions, clicks, unique visitors, total visits, and page impressions. These metrics, all measurable with server logs, IP addresses, and cookies, form the backbone of the online advertising industry. The PWC report, drawn from many of the largest corporations involved in online advertising, provides a detailed map of the advertising process, complete with standardized metrics and reporting. The IAB, the AAAA, and the ANA quickly adopted PWC's recommendations into an industry-wide standard in 2002.[55]

Although the IAB succeeded in standardizing metrics and in collaborating with other standards consortia, competition among marketers did not end; it simply shifted to competition for customers and the creation of new techniques. For example, jumping forward in time to today, the IAB counts as members putative competitors, such as Microsoft, Google, Facebook, Twitter, Yahoo!, Digg, eBay, and Amazon.[56] While news reporting on these companies often presents them as locked in life-or-death battles for Internet commerce,[57] at IAB meetings, their representatives collaborate to create and maintain marketing standards, and they speak with a collective voice to regulators and other industries. By coming together, competitors have largely solved the "user problem"—the chaos of the Web and its impact on advertisers. By roughly 2003, with the advent of broadband and mobile networks, and as more and more people began using the Web, a new, contained, predictable advertising market was born. This is the market in which social media arose.

Social Media as a Standardized Market

To explore social media as a product of IAB advertising standards, we can start with information architecture. We know that the vast majority of content in social media is user-produced. However, the architecture of a social media site does not just include the content implemented by users; it also includes spaces for visible advertisements. User content within the social media frame is thus mixed together with marketing. This process received unique praise from Tim O'Reilly in his defining document on Web 2.0: "People don't often think of it as 'web services,' but in fact, ad serving was the first widely deployed web service, and the first widely deployed 'mashup' (to use another term that has gained currency of late). Every banner ad is served as a seamless cooperation between two websites, delivering an integrated page to a reader on yet another computer."[58] In other words, a website is a "mash-up" of top-down, incrementally altered architecture, bottom-up user participation and processing, and the lateral insertion of advertising, creating a coherent visual artifact out of these different streams. This mash-up is made possible with standardized protocols, including those developed by the IAB.

This nexus of architecture, user-led content creation, and advertising is our window into an early competitive development in online advertising standardization: ad networks. Whereas in the late 1990s online advertising largely followed a mass media "portal" model, where a site such as Yahoo! would produce editorial content and services to attract a large audience and then would sell advertising space around that content, the Web grew to include billions of small, niche sites: home pages, discussion boards, and Web rings. Ad networks were formed to reach these small sites (and the audiences they were building). The advertising company would create advertisements and then contract with an ad network that could distribute them across the Web. That way, the advertiser would not need to search for and contact thousands of small-site operators; the ad network handled that work. Today, Google AdSense is perhaps the best known ad network, although Yahoo!, Microsoft, and AOL are major players in ad networks.[59] Their networks were built on the standards created by such organizations as the IAB. Rather than negotiate heterogeneous sizes with these thousands of small sites, ad networks could insist that small websites be built around one or two standard ad sizes.

Ad networks were thus an early success story in "monetizing" user-generated content. Rather than use the mass media model (as in the Yahoo! portal with professionally created content, or as in the *New York Times* model of journalism surrounded by ads), ad networks united amateur content creators (blog owners, for example) with major brands and marketers. As ad networks gained traction in the 2000s, so did social media darlings, such as blogs and wikis; their user-led rise was celebrated at the same time that advertisers had figured out ways to spread marketing messages to blog and wiki niche audiences.

Of course, ad sizes were not the only standards that allowed ad networks to function. The standardization of metrics, terminology, and legal interactions produced by the IAB and its allies set the stage for competition and experimentation at the data-collection level behind the surfaces of social media sites.[60] The network effects of the IAB's standardization includes ad networks developing state-of-the-art tracking software, consumer profiling, and cross-platform data sharing: "Leading marketers are . . . incorporating new tools, such as targeting based on consumer behavior, database marketing, Web analytics, and predictive modeling, in planning their media activities.

These are the companies that stand the best chance of 'torturing the data until it confesses,' as Paul Price, global president of ad agency Rapp Collins, puts it."[61] As ad networks insinuated themselves into niche sites, marketers began to learn more about the desires of Web users. Surveillance of user activities has become the business model of online advertising.[62] As Christopher Vollmer puts it in a report sponsored by the IAB, "To 'activate' a consumer [i.e., bring a consumer into relationship with a brand], you must understand the consumer. To understand a consumer, you must listen and observe. When you listen and observe, you drive insights."[63] He links online surveillance to previous forms, such as monitoring consumers in shopping markets and in doctor's offices. Of course, the possibilities for monitoring users on the Internet far exceed those of "real space."

A variety of techniques allows websites and advertisers to "listen and observe" users while they are online. Cookies, of course, are well known; these are small text files that sites place on a user's computer to identify the user's browser, computer operating system, IP address, and (if the user provides it) personal information. Somewhat less well known are "Beacons" (LSOs, sometimes called "Web bugs"), 1x1-pixel invisible GIFs that allow sites to transfer cookie data to third-party ad networks and analytics firms. This traffic allows ad networks to track users across multiple sites; thus, even users who do not provide personal information will have individuating profiles built out of their online activities.[64] Cookies, of course, can be blocked with modern browsers, and they are limited to a size of 4K. Less often blocked (because they are not as well known) and less constrained in terms of size, Local Stored Objects (LSOs, sometimes colloquially called "Flash cookies" or "super cookies") built on Adobe's Flash technology allow marketers to track mouse movements (as a proxy for eye movement) and time "interacting" with an advertisement (for example, moving an animated object, playing a video file, or clicking a link). Much like the flight-path data that Norbert Weiner worked with in his attempt to create cybernetic antiaircraft gun systems,[65] the tracing of mouse paths across Flash objects illuminates the likely path that new visitors will take. This information is stored in "Flash cookies" that can hold up to 100K of data without user intervention and are stored outside the browser (usually in a folder installed on the user's computer when he or she adds Adobe Flash). Users do not often know which applications are using them.[66] Like their cookie forebears, Flash cookies can be linked within ad networks for cross-site tracking of the IAB standardized metrics: unique visitors, impressions, clicks, and interactions.[67] With traditional or Flash cookies, only a small amount of data needs to be gathered to precisely identify and track a user's computer.[68] Once the user is tracked, a behavioral profile can be built and used to more precisely target ads to the user's desires.

However, possibly the greatest expressions of the IAB's marketing standards are such sites as Facebook, Google, and Twitter. Their goal, from their earliest days, has been to produce audiences that adhere to tracking standards and deliver them to marketers.

For example, early social networking site Friendster sought to make explicit the implicit social connections found on the Web.[69] Social networking sites require users to build profiles of themselves, consisting of predetermined demographics (such as sex, age, and hometown) and interests. Although practices have varied across different social networks, such sites as Facebook and LinkedIn have in part transposed "real-world" social networks of friends to the Web. Facebook's roots in colleges has also contributed to the use of "real-world" identities online. These practices have resulted in a rich dataset of connections, complete with the socially articulated desires of specific individuals. *These are, in fact, precisely the IAB-standard datasets that marketers have worked to build via online tracking and surveillance.* The only difference, really, is that in the case of social sites, the users do much of the work of profiling themselves. Users labor to produce themselves through declaring their desires, making themselves desired objects of the marketing trade.

As social media allow us to be more open about our desires, we produce our own ontologies and metadata on such sites as Facebook: Sam is a friend of Sue; Sue is a fan of BMW. When Sue "likes" the latest BMW model, her desire is pushed out into the social stream.[70] Thus, Sam sees Sue's earnest expression of desire not as an advertisement or not even as cybernetic metadata about Sue, but as part of Sue's overall subjectivity; Sue truly is a fan of BMW, after all. This is a seamless Web connecting personal metadata to many people and brands. Its ontological politics include the flattening of friendship into equally weighted relationships distributed among flesh-and-blood people, socialbot avatars, and fetishized, ephemeral brands.[71]

This "Semantic Web" of desire produces finely tuned archives of affect: "Due to the extensive collaborations online many applications have *access to significantly more metadata about the users*. Information about the choices, preferences, tastes and social networks of users means that the new breed of applications are able to build on much richer user profiles."[72] Linking users and brands in a semantic social graph allows marketers to weave their messages into the daily conversations and browsing habits of users. A *New York Times* article on IAB member Acxiom (a data-mining and marketing company) describes a scenario where a user, Scott Hughes,[73] sees on Facebook that his friend is now a fan of Bryce Computers:

> When Mr. Hughes follows a link [from Facebook] to Bryce's retail site, for example, the system recognizes him from his Facebook activity and shows him a printer to match his interest. He registers on the site, but doesn't buy the printer right away, so the system tracks him online. Lo and behold, the next morning, while he scans baseball news on ESPN.com, an ad for the printer pops up again.[74]

The link between friend and brand is compelling. As the IAB's *Social Advertising Best Practices* manual explains, "The ad experience can become even more relevant by virtue of a consumer endorsement in the ad itself. A friend's photo and explicit endorsement or explicitly shared information about how the friend has interacted with the [brand, product, or store] can be displayed in the ad, making the ad personally relevant to an unprecedented degree."[75]

To be fair, for much of the 2000s, social networking sites were (to the dismay of marketers) "walled gardens" with built-in limitations to third-party access.[76] Facebook and Myspace, the two dominant competitors for much of the decade, were even so incompatible with one another that very different user bases gravitated to each; typically,

Facebook users were better educated and more affluent than their Myspace counterparts.[77] However, once Facebook achieved a significant market share, the advent of the Facebook Connect protocol broke user data out of the "walled garden" and allowed third parties to build applications that could be interposed between the user and his or her Facebook "social graph" of friends. Sites as various as the *New York Times* and CNN to putative competitors YouTube (owned, after all, by Google) and Myspace were linked to Facebook via Connect. Facebook Connect is the ultimate expression of the standards-setting project of the IAB; after spending years building up a user base via network effects, Facebook's IAB-inspired standardized datasets were opened up to marketers across the Web.[78]

Thus, social media templates have developed in large part as a result of the standardization of advertising practices established by the IAB. The "user-led" Web has been structured from its earliest days as a protocol for capturing and storing the user affect. Essentially, advertisers stopped buying space on websites and started buying access to the hopes and desires of users. Investors' valuation of such data is staggering: Google's purchase of user-led media sites, such as YouTube (for $1.65 billion), links with its purchase of ad networks DoubleClick (for $3.1 billion) and Applied Semantics (now called AdSense, purchased for an undisclosed amount). YouTube allows users to create content and thus audiences; DoubleClick and AdSense allow marketers to reach those audiences. Microsoft's investment in Facebook (for $240 million in preferred stock, at the time valuing Facebook at $15 billion) links Facebook's "social graph" to Bing searches and Facebook's user data to Microsoft's advertising networks.[79] Framing user-led production within a standardized template derived from a seedbed of surveillance standards has been lucrative.

Conclusion: The IAB's Self-Regulation, Its "Cerned" Subject, and Avenues of Resistance

Of course, the methodical transformation of the Web from a non-profit space for academic and intellectual exchange to yet another standardized, template-driven, advertising-centric medium has not gone unchallenged. Government regulators, such as the U.S. Federal Trade Commission, Canada's Privacy Commissioner, and the European Parliament, have to varying degrees sought to restrain Web-based companies from collecting and retaining data on their users. Much more vocally, privacy advocates, such as the Center for Digital Democracy and Privacy International, have consistently argued against the expanding tracking and surveillance conducted by online marketers.[80] Thus, throughout its history, the IAB could not ignore politics and focus solely on producing advertising standards on the Web; like any other private standards consortium, it also had to act as a trade organization and lobby against these challenges. In doing so, it has deployed the third tactic of contemporary standards consortia: the ideology of self-regulation.

This ideology relies on the production of what I call (following Paul Smith) the (dis)cerned theoretical subject of the IAB: the *sovereign interactive consumer*. Smith argues that every discipline—and, of course, marketing is an accepted discipline with its own theories and methods—catches subjects in a "mastering theory" that delimits them, a process he calls "cerning." This subject is an abstraction, "removed almost entirely from the political and ethical realities in which human agents actually live."[81] To achieve this abstract state, the sovereign consumer must "first accept a given inheritance or patrimony . . . and, second, be encircled or surrounded, in a synecdochal figure or by the definition of the whole in and through which they accede to 'real' existence."[82] The IAB in particular, and marketing in general, produces such a subject through theories of how humans live.

Like any good theory, marketing's theory must reduce the chaos and unpredictability of humanity to a rationalized, predictable shape.[83] For all the protestations about the Web being user-led, marketing as a social science must produce an object (in this case, a particular subject—the sovereign interactive consumer) that can be studied scientifically. As I argued in Chapter 3, although this subject is abstracted from concrete, human reality, it can become a real abstraction, a subject position that is filled by concrete people. The configurations and scripts people are invited to follow can be compelling, and the limitations imposed on them can be strong enough to modulate behavior and thus give the abstract theory some concrete life.

There is a contradiction here. This delimited and constrained subject is produced through the IAB's insistence that the subject is, in fact, powerful, self-producing, and autonomous. The IAB's subject inherits decades of neoliberal ideologies of self-determination and personal risk management.[84] Hence, the sovereign interactive consumer is a free subject, able to choose whether to visit advertising-supported media, fill in a social media profile, take on the risks associated with placing personal data online, or become educated about behavioral advertising practices. The sovereign interactive consumer equates consumption with freedom and as such takes *caveat emptor* to heart as he or she chooses which social network to sign up for. He or she is not a laborer; the sovereign interactive consumer is a producer (or produser?),[85] but he or she does not "work." Instead, the sovereign interactive consumer self-produces via the playwork of ludic interactions with digital artifacts, such as chunks of data (text, images, videos, audio, etc.) or fetishes (avatars, profiles, user names, logos, brands, etc.). Above all, the sovereign interactive consumer is the master of digital flows, first of the selection and control of informational flows he or she receives (particularly information about the efficacy of various products and services) and next of outward flows: affection toward various digital objects and digital payments that realize the value locked inside a wealth of consumer goods. In publicly performing this role, this subject coheres into an individual, gaining power by building a personal brand.[86] Indeed, "the promise of interactivity," writes Mark Andrejevic, "is that viewers can be producers as well as consumers— that, furthermore, their participatory consumption can be creative and fulfilling."[87] Thus, the product of such interaction is the sovereign interactive consumer's own increase in pleasure, convenience, and power; as global corporations alter their advertisements and product offerings to meet the consumer's desires, this individual shapes the world. For the mastering theory of marketing, this powerful personification is the foundational abstraction of social media.

To combat the arguments of privacy advocates, the IAB (along with several other marketing trade groups)[88] has produced this subject as well as the documents guiding its interaction with it: the "Self-Regulatory Principles for Online Behavioral Advertising," a work that centers on such ideals as transparency, accountability, and education.[89] Because the sovereign interactive consumer is so powerful, the IAB's only responsibility is to provide information, such as privacy notices about online marketing and profiling. Indeed, in the best tradition of corporate education, this information is provided via the website Privacy Matters, linked through the IAB home page.[90] There, a visitor to the site can click through Javascript- and Flash-powered factoids about cookies and tracking. Thus, if the sovereign interactive consumer *chooses* to be educated, to understand the workings and benefits of behavioral advertising, the IAB is ready. If not, so be it.

If, perhaps, the sovereign interactive consumer does not realize that this interactive and engaging educational resource exists, he or she can find similar information by clicking a tiny triangle logo located on some advertisements to go to another educational site and learn about how that particular behavioral ad got there.[91] Beyond this, the "Self-Regulatory Principles" also require adherents to provide "opt-out" cookies, available at AboutAds.info.[92] Visitors to that site have their computer scanned for about a minute, and then they may select cookies from more than one hundred advertising companies participating in the program.[93] Once this is completed, the user's browser is inundated with these cookies. Those users who regularly delete their cookies must return to the site and redo the process to get them back—that is, unless they set exceptions in their browser for the more than one hundred different opt-out cookies. However, it takes a keen eye to tell an opt-out cookie from one being used to track Web activity.

Finally, for the most active sovereign consumers, at that same AboutAds site,[94] users are allowed to fill out a form to report to either the Better Business Bureau or the Direct Marketing Association any ad companies that violate the opt-out cookies or otherwise act unethically. One wonders how the sovereign consumer would be able to tell whether he or she were being tracked despite having opt-out cookies from a bewildering array of advertising firms or whether these ads are simply serendipitous. But, of course, in the abstract theory of the

IAB, the sovereign interactive consumer is capable of managing all this as well.

To sum all this up: because the sovereign consumer is powerful and self-interested, he or she is able and welcome to assume the risks for posting personal data online, study factoids about behavioral marketing, watch out for a tiny triangle logo, click through page after page of inscrutable opt-out information and disclosure statements, customize opt-out cookies in his or her browser, monitor the voluntary compliance of the IAB's hundreds of members IAB, be wary of the countless tracking companies that *are not* voluntary members of the IAB, and read corporate privacy policies, press releases, and, of course, the "Self-Regulatory Principles" document itself. These are the sovereign consumer's rights and responsibilities. If he or she chooses to ignore these in favor of other activities (such as the pleasures of using social media), the consequences of having his or her Web use tracked, of having websites hide or highlight certain information to prompt specific responses, or of having credit scores or employment options affected are entirely the sovereign consumer's fault.

As Richard Maxwell and Toby Miller argue, "The symbolic power of media technology is enhanced by the idea of the liberated consumer, which, like the commodity sign, provides no residual correspondence to a reality other than its own."[95] This symbolic power, enhanced by the abstract sovereign consumer, shapes the online advertising industry's relationship with regulators. As pathetically anemic, inconvenient, and ineffective as these self-regulatory efforts are, for the most part the IAB has convinced government agencies that its imagined subject, the sovereign interactive consumer, is equipped to deal with massive, ubiquitous advertising networks; the complex interaction between first-party and third-party Web analytics; myriad flows of personal data; and, of course, the sophisticated noopower of advertisements that plays on fears, memories, and desires (a power developed over nearly a century of intensive and expensive neurological, psychological, and sociological research). With the fiction of consumer sovereignty produced, the IAB has a potent argument for self-regulation[96]—and it works.[97] In the United States, for example, the Federal Trade Commission (FTC) and Barack Obama's administration have both endorsed self-regulation as the best mechanism for regulating online marketing.[98] Thus, the IAB's efforts to build a particular subject and sell the idea of the existence of that subject to government agencies has been largely successful.

Of course, turning to theories of how protocols and standards work, we see that the IAB's multiple standardizations (both of the technology and of a desired subject) are not benign appeals to our individual, sovereign power. Rather, such protocols regulate behavior, delimiting its contours and subtly determining its direction. We are never sovereign. Our language and actions are modulated. As Alexander Galloway and Eugene Thacker note, "Protocological networks are inclusive rather than exclusive; discrimination, regulation, and segregation of agents happen on the inside of protocological systems (not by the selective extension or rejection of network memberships to those agents)."[99] That is, these systems do not exercise power by preventing Person A from joining the network while allowing Person B to do so; rather, they draw objects into the network so long as those objects "speak" the protocological language. As Facebook Connect identification, Google search functions, and YouTube, Scribd, and Flickr content management systems link up via APIs, ad networks, and acquisitions and mergers, they produce what I have elsewhere called the "Web 2.0 Portal."[100] We are increasingly drawn into networked social media capitalism, and within these boundaries we are typified, aggregated, and regulated. This is noopower in action, acting on us before we even load the next page in our browsers. It should be clear at this point which language we are invited to speak in this system, because we agree to it when we agree to explicit (if often buried and jargon-infested) terms of service statements and privacy policies as well as implicit protocological structures, such as the cultural practices of friending, following, liking, and linking. We are invited to speak the language of consumption and little else. And we agree to speak it publicly for all who would listen.

While all of this sounds quite bleak, standardization produces its own forms of resistance. Technical standards have weak points that human beings can and do exploit and alter. The practice of exploiting and resisting marketing while benefiting from the "free lunch" that advertisements subsidize is a quotidian practice of humans living within informational capitalism, one that does not square with the consumer-subject whom the FTC and the IAB imagine. As Galloway has argued, "The contradiction at the heart of protocol is that it has to standardize in order to liberate."[101] This contradiction is a key hole (to use the language of information technology and software security)[102] in the structure the IAB has created. Standardization can, in fact, bring about liberation, but only when standardization is exploited for political purposes.

There are several examples of the ways in which end users can exploit standardization. On a basic level, the IAB standards cannot prevent polysemic games played with the ads. Users can appropriate them, culture-jam them, and digitally alter them. This form of resistance to advertising has been the staple subject of much media studies. However, it is difficult to see this activity as truly resisting the power of advertising standards, and it is commonly appropriated by the creative people behind marketing. As Jack Bratich argues, "Sniping from an 'outside' (a la culture jamming) might be obsolete in a situation where real subsumption has made the marginal central to production."[103] Moreover, as advertising networks continue to refine and target ads, culture-jamming loses its mass media referential power. Digital advertising is now engaged with extremely refined experimentation based on behavioral tracking: the ad I see may be different from the one you see, diffusing the common ground of references and signs required for a culture-jamming reappropriation.[104] In other words, my culture-jammed advertisement may be so radically different from the personalized ads you see that the semiotics of my culture-jam may be lost on you (and vice versa).

More powerfully, standards allow for ad-blocking software to work. Browser extensions, such as the extremely popular AdBlock Plus (for Firefox, Safari, Opera, and Chrome), Simple Adblock (Internet Explorer), and NoScript (Firefox), can block ads from being displayed on a user's computer. This is possible because the distribution and placement (if not the content) of advertising is so standardized: as advertising networks congeal and as social network interfaces are built around space to be sold for advertisements, creators and users of ad-blocking software have an increasingly easy time preventing browsers from rendering ads. They can block known ad network URLs, Javascript, Flash, and commonly used locations within the social media template. Similarly, the growing push for a "Do Not Track" option in browsers would be successful only if standards consortia, such as the IAB and the W3C, could agree on standards for compliance. In these cases, standards can be used against the very industry that sets them. However, these practices can be appropriated and resisted by the industry. Version 2.0 of AdBlock Plus now defaults to allow "acceptable ads." The definition of "acceptable" is unclear, but what is clear is that this definition will be influenced in part by the advertising industry.[105] The online advertising industry even offers its own ad-blocking program, Ghostery, which helps marketers see where their ads are appearing and how often they are blocked.[106] Finally, IAB members may decide to ignore "Do Not Track," since they are self-regulating and no state-based regulation requires them to obey.

In the end, perhaps the most powerful move to resist online behavioral tracking will arise much in the way that Karl Marx argues socialism will arise from capitalism: like the organizing power of capitalism, which brings workers together in locations where they can discuss their common grievances and begin to organize, if social media protocols provide us with a common language, we can use that language to discuss our common problems and find ways to modify social media software. As Galloway and Thacker note, "Computer viruses thrive in environments that have low levels of diversity. Whenever a technology has a monopoly, you will find viruses. They take advantage of technical standardization to propagate themselves through the network."[107] Like a computer virus, discontent and a desire for something better can spread quickly in standardized systems. The better the IAB and its progeny (social networking sites, the search monopoly Google, advertising networks, and major content providers, such as traditional media companies) get at standardizing the shape of the Web, the more open the Web is to being hacked, altered, and shaped by resistant actors who increasingly see their common cause. Even the *Wall Street Journal*, no flag-bearer of progressive politics, has done in-depth investigations into the power of the online advertising industry.[108] The fatigue of being tracked is apparent among people across the ideological spectrum. Much as the organization of the factory necessarily meant the organization of labor, which led to the rise of unions and solidarity among workers, the aggregation of free-laborer users together within networks in which communication is enhanced can lead to potent anticapitalist activism. Chapters 5 and 6 explore this concept in more detail.

5

Engineering a Class for Itself

The Case of Wikipedia's Spanish Fork Labor Strike

> Revolutionaries see history as a creation of their own
> spirit, as being made up of a continuous series of violent
> tugs at the other forces of society—both active and
> passive, and they prepare the maximum of favorable
> conditions for the definitive tug.
>
> —Antonio Gramsci, *Selections from the Prison Notebooks*

So far, this book has largely been a response to the very valuable analyses of networked labor put forward by such scholars as Tiziana Terranova;[1] Hector Postigo;[2] Detlev Zwick, Samuel K. Bonsu, and Aron Darmody;[3] and Mark Andrejevic.[4] Specifically, Terranova's seminal essay "Free Labor," published in 2000, provides a very clear anticipation of the contemporary social media processes explored in this book. As she explains, "Simultaneously voluntarily given and unwaged, enjoyed and exploited, free labor on the Net includes the activity of building Web sites, modifying software packages, reading and participating in mailing lists, and building virtual spaces on MUDs and MOOs."[5] In other words, free labor is a highly contradictory process. Besides being freely given yet exploited, such work is often compelling and yet exhausting: the constant flow of social media streams invite us to watch, update, and participate, but these flows, part of so-called push media, never end and thus constantly demand our attention. Free labor is an act of love, but it is also reducible to the cold logic of exchange: we make and maintain relationships on Facebook and YouTube, and the emotions associated with those relationships are powerful, and yet in the end these linkages become reduced to nodes in a social graph. Free labor is also rarely valorized, but when it is, it is hypervalorized: most users do not get a financial reward for their participation, and yet, of course, some become microcelebrities[6] and build followings that can be traded for cash.[7] Finally, free labor is highly pervasive and endemic to late capitalism, and yet it is often perceived to be best performed by talented individuals. In other words, we are all invited to contribute images, texts, and ideas to social media, but many of us think of this work as amateur in comparison to the creative laborers closer to the center of traditional, mass media production; thus, we undervalue the creative work we do within social media, even as it becomes hypervalorized in stock-market IPOs. Regardless of how it is valued by the producer, both amateur and professionally produced flows of information are essential to late capitalism.[8]

Indeed, this book has explored this idea at length. The future, we are told, is user-led and driven by the freely given affective work of users. But who has established the road the user leads us down? And who has determined the destination to which we are headed? Or are we simply going in circles, with no clear leader and no end in sight?

The analyses of free labor tend toward the latter conclusion—that is, there is something circular or circumscribed about user-led production in social media as in other places. In so many ways, there are dialectical relationships between structure and agency, work and play, and self-production and socialization. We are told again and again that this new way of work, this new digital economy, is revolutionary, because power has shifted to subjects like the user and the consumer. But circles and loops are revolutionary only insofar as they revolve.[9] In this case, the processes of production, distribution, and consumption—as well as those of valorization, commodification, and exploitation—continue to revolve around a fixed axis, and the kinetic power generated by such revolutions flows directly to new media capitalism.

The contradiction maintaining this circular process has been explored at length in the Autonomist Marxist school by such theorists as Michael Hardt and Antonio Negri.[10] In their work, they argue that free labor (or, as they call it, biopolitical production) is a way for labor to become autonomous from capitalism. Because so much free labor involves acts of love and freely given gifts, we can imagine that those engaged in this process can see that exploitative capitalist production is unnecessary. After all, the profit motive does not explicitly enter into the desire to give away one's labor, or at least it is often deferred. And yet, while waiting for this gift economy to slough off capitalism, practitioners of free labor see their work eagerly vacuumed up by networked capitalism. In the end, it may be the promise of liberation (i.e., a promise of a gift-based economy), or perhaps, if we are being more cynical, the promise of hypervalorization or celebrity (i.e., effects of a profit-driven economy) holding the system together and endlessly deferring alternatives to networked capitalism.

The work of critical theory is to find a way out of this infinite loop. Unfortunately, much of the work on free labor fails to do this. Instead, studies of free labor opt for detailed analyses of the revolutionary movements of capital from production to consumption and back again. Indeed, a criticism leveled at Autonomist Marxist analysis accuses it of holding onto a faith that the circular flows of value and information will be broken by an immanent rupture of the multitude.[11] Like the action-crushing sense of the inevitable socialist revolution (attributed to Karl Marx's theory of the falling rate of profit), or perhaps like the Day of Reckoning that is, for certain evangelical sects of Christians, announced on highway billboards to be always a few months away, we wait for such a rupture. And wait. And wait. And wait.

For better or for worse, I want to avoid this endless loop. To do so, I want to point to a moment in the history of social media when such a loop has been broken, and this is found in the case of Wikipedia and what I call the Spanish Fork Labor Strike. I call this incident a "labor strike" because this is an instance when free laborers (largely located in Spain) recognized their status as such and withheld their labor to have greater say in the shape of the nascent Wikipedia. In 2002, in response to the possible inclusion of advertisements on Wikipedia.com, the strikers stopped working for Wikipedia and instead set up their own alternative, open encyclopedia, *La Enciclopedia Libre Universal en Español*. For a period of time, *La Enciclopedia Libre* grew faster than the Spanish-language Wikipedia. The strike forced Wikipedia founders Jimmy Wales and Larry Sanger to change many of the policies of the nascent Wikipedia. In many ways, Wikipedia has its current shape as a nonprofit, marketing- and (top-down) surveillance-free space of freely given labor because of the Spanish Fork Labor Strike. In other words, this strike is why we visit www.wikipedia.org instead of www.wikipedia.com.

But ultimately, as much as I want to, I cannot entirely attribute Wikipedia's more egalitarian shape to the labor strike. To be sure, I desperately want to have this simple causality: labor strike leads to nonprofit status leads to less exploitation of users. After all, if it were so simple, I could suggest that we could simply stop using Facebook or Google until it meets similar demands.

This is where the software studies orientation of the book comes into play. What made the Spanish Fork Labor Strike possible was not just the will of the Wikipedia participants in Spain; it was also several software architecture decisions made by Wales and Sanger. These decisions, which included the use of the GNU Free Documentation License (GFDL) and UseModWiki software, provided technical affordances to the Spanish strikers. In any story of the Wikipedia labor strike, we have to consider the case as an overdetermination among structure and agency, software and users, love and labor, architecture and implementation—indeed, the very dialectics brought to light by analyses of free labor.

Thus, in the end, this chapter is less about Wikipedia and more about the heterogeneous engineering of a particular social formation: a *user class for itself*, one produced out of the *user class in itself*. It is about the mix of technical and discursive artifacts that such a class comprises as well as the sheer *work* of producing such a class. This is, I think, the element largely missing in the theories of Autonomist Marxism and critiques of free labor: multitudes do not erupt spontaneously. They must be engineered into being by someone, somewhere, who is offering up free labor to bring a class for itself into being.

"Class for Itself" as a Discursive-Technical System

Beginning in Chapter 1, I have taken for granted the idea that the users producing content within the social media frames are a class of free laborers. I have pointed out software architectures of social media and how they help shape uses, and therefore users, of social media. Given that users' labor is geared toward processing the new, that the products of this processing are alienated from them (via Terms of Service agreements) and stored in rationalized archives, and that interactions within social media are shaped via templates and protocological containers to better produce users as interactive consumers, it is clear that users relate to social media as laborers. Following Steve Woolgar, we can say that the material and discursive contours of social media help "configure" the user in this way.[12] In Woolgar's sense, the engineering of a technical artifact is in part a process of a priori defining, enabling, and constraining users; users are imagined early on in technical engineering as having certain capabilities and desires. Engineering then becomes a process of negotiation with this imagined user: some imagined user capabilities are taken advantage of, while others are extended; some imagined desires are met, while others are cordoned off and denied (or "effectively frustrated," to use Tarleton Gillespie's term).[13] In the case of social media, site designer's goals from the outset were to harness what Eran Fisher calls the "people power" of Web users.[14]

Thus, the user of social media is, in fact, *engineered* from heterogeneous bits and pieces, including the technical structure of networks; the deployment of such terms as "participation," "platforms," and "social"; Western discourses of the rational, coherent subject; and histories of understanding communication as a normative practice. As I argued in Chapter 3, as users implement their part of the architecture—that is, as they build content, make connections, and process digital objects—users *themselves* are being abstracted. Social media abstractions are not only a means for software architects to think through the organization of software-on-hardware—that is, as a matter of architecture and implementation (and the associated management of labor); they are also a social force, a "real abstraction" or analytical category that is capable of producing hierarchical, subjective experiences of computer use.[15] As Enzo Paci argues, "The fundamental character of capitalism . . . is revealed in the tendency to make abstract categories live as though they were concrete."[16] That is, capitalism functions to make each of us live out abstract categories (laborer, owner, member of the middle class, lawyer, investor, coder, computer user) as if they were concrete, embodied, and real. "In the final analysis," writes Alberto Toscano, "something really happens when abstraction takes place. Abstraction transforms (and the fact that what it transforms is itself abstract does not make it any less real)."[17]

Given the material and technological power of social networks, the architecture can call forth a subject even before the subject learns how the system works. The process of real abstraction works because the practice of creating abstractions (such as in the practice of abstraction in software engineering—see Chapter 3) produces categories of thought and thus subjectivity among those who in turn work with abstractions. Real abstraction is noopolitical; it is "thought before thought." Processes of division and generalization create thoughts in us, and we reinforce those thoughts by continuing processes of division and generalization.[18] The division of labor is an example: breaking production down into smaller tasks produces the thought of the division of labor, a thought that has material effects in that it can produce workers who *think of themselves* in terms of their tasks and their tools. As Anni Dugdale puts it, "It is the mixing together of such materials with bodies that constitutes subjects of a particular kind. For the subjectivities of the participants are already being produced in these material arrangements, even before any verbal performances have occurred."[19] That is, before anyone explains to a new Twitter user how Twitter "works," the new user confronts a real software abstraction, a framework that suggests certain uses and subtly denies others. The shapes of Facebook or Google+ are not accidental but are built with (idealized) uses in mind. Likewise, the (ideal) user must be socially engineered. The process of producing such a user is integral to the production of social media architectures themselves.

If the user is configured to live out the abstract category "free laborer," then it is clear that the mass of users has been engineered to be a "class in itself," to use Marx's term. As he argues in an analysis of the birth of classes in capitalism, "Economic conditions had first transformed the mass of the people of the country into workers. The combination of capital has created for this mass a common situation, common interests. This mass is thus already a class as against capital, but not yet for itself."[20] As in the transition from feudalism to capitalism, contemporary social media capitalism has a massifying effect. The repeated emphasis in IPO-filing statements and tech reporting on numbers of active users and graphs of flows of traffic reveals the necessity of a mass of users to make social media function.[21] Social media requires a mass of user free laborers.

It appears ridiculous to say that social media users are a "class against capital"—that is, against social media capitalism. Social media, for all their *social* elements, are oddly individuating phenomena. Participants on LinkedIn, Twitter, and individual blogs often work to build their own personal brands or statuses as microcelebrities. This appeal of individual empowerment is a central feature of the Web 2.0 ideology of the erosion of centralized gatekeepers and mass media power: social media are democratizing forces capable of flattening hierarchies and putting individual users on equal footing with experts, big corporations, and governments. This is, in sum, liberation from the deadening bureaucratization and oppression of mass culture, a liberation of the sovereign individual. And the unique individual is never a member of a class; the user rather is a singularity, a monad, an irreducible collection of personalized tastes and desires. "There is no such thing as society," Margaret Thatcher famously told us. And yet, to borrow a sentence from Marx, "In spite of manuals and utopias, combination has not yet ceased for an instant to go forward and grow with the development and growth of modern industry."[22] That is, despite the promises of social media—promises of equality, of the eradication of gatekeepers, of media empires brought down by tweets, and of personalization and individuation above all—class is *engineered* into social media. It is part of its constitution; each individual, sovereign user participates in a system largely interested in *aggregated* labor and the *mass* production of desires and affect. Moreover, the appropriation of the free labor of users is a mass process; data are aggregated into patterns of affect, and users are typified into various consumer categories. As I argued in Chapter 1, this massification is precisely why socialbots are even possible.

If this is the case, to follow the normative Marxian path, we have to consider how a class *in* itself becomes a class *for* itself, capable of articulating its interests and resisting exploitation. The classical Marxist response is to point to the *inevitable* subjective consciousness that arises among a class in itself as its members realize their shared, objective plight (arising out of the *inevitable* immiseration of workers); this understanding *always* leads toward class for itself. However, across the literature on class for itself is a denial of the inevitability thesis: that is, historical and sociological studies have shown that class consciousness does not inevitably arise as capitalism polarizes people into the two grand competing groups, owners and workers.[23] Indeed, even this polarization is thrown into doubt with new sociological groupings, such as the middle class, the "Professional Managerial Class," creative/knowledge workers, and the personally branded, all muddying the supposedly neat bifurcation of society within capitalism. Again, the revolution is endlessly deferred as we wait for the proper polarization to appear.

Instead, I suggest that the production of a class for itself mirrors the engineering of a social media class in itself. It is decidedly difficult, associative work that needs to be conducted along multiple fronts: politics, culture, economics, and technology. Indeed, as Michael Buroway puts it, in a slightly different context, "Consciousness and interests do not necessarily coincide."[24] They have to be made to coincide. This is, to borrow from Antonio Gramsci, a matter of a "war of positions" across cultural, economic, and political spheres. This denies the inevitability thesis for something I think is much healthier: the need for critical activism that takes nothing for granted and does the difficult work of crossing social boundaries to articulate commonalities among groups of people. As Stuart Hall argues (drawing on Gramsci):

> The important thing here is that so-called "class unity" is never assumed a priori. It is understood that classes, while sharing certain common conditions of existence, are also cross-cut by conflicting interests, historically segmented and fragmented in this actual course of historical formation. Thus the "unity" of classes is necessarily complex and has to be produced—constructed, created—as a result of specific economic, political and ideological practices. It can never be taken as automatic or "given."[25]

In short, class for itself becomes a problem for activists engaged in heterogeneous engineering, just as a class in itself is engineered by the architects of social media.

This engineering occurs in three steps. First, exploitation must be articulated. The burden is on those who want to form a class for itself to prove that (1) the owners of productive assets are gaining wealth and/or power while the nonowners are being deprived, (2) this gain arises because the owning class excludes nonowners from access rights to productive resources, and (3) the owners appropriate a surplus from the productive activities of nonowners.[26] To prove these assertions, and thinking specifically of this in relation to social media, members of a would-be class of users must associate discourses (such as Terms of Service agreements, intellectual property regimes, definitions of value, concepts of fairness, transparency, participation, etc.), and technologies (systems of exclusion and intrusion detection, systems of punishment for violations, mechanisms of appropriation, new capabilities gained from the purchases of start-ups, etc.) into a complex political economic picture that clearly demonstrates the existence of exploitation. This picture also bears the burden of being opposed by alternative articulations of social media (i.e., social media owners are rich not because of appropriation and exclusion but because they are technical and business geniuses; we are not being exploited in social media because we are getting so much back from these sites for free; no one forces us to use these services; we enjoy them; etc.).

Next, if elements are associated in such a way that exploitation is demonstrated, discursive and technical pressure points in the system must be produced. A very simple pressure point is that *social media site owners need users.*[27] Specifically, they need them to do the work of digital and affective processing within a standardized frame. This effort can be withheld.

Finally, perhaps the most important articulation is that of alternatives. Even if the members of a nascent class for itself recognize the sheer dependence that social media site owners have on them, with no other outlet for the very real pleasures of social media, they are likely to stay put, even in exploitative conditions, and particularly if the sites remain "free" in the sense that they are free of charge. The network effects of exploitative services are powerful enough to keep a critical mass of people who can do a cost-benefit analysis: the service I get for my private data is payment enough—how else could I stay connected?

This mirrored process of engineering class for itself is not just the purview of activists. Social media capitalists recognize these elements as well. Site owners must actively construct systems that can most effectively exploit user desires; pressure points and weaknesses are recognized and shored up (very often through iterative alterations in both legal and software code—an excellent example is Facebook's Terms of Service, which has been altered many times over the years to appropriate user content while denying its intention to do so); and alternatives must be appropriated (start-ups get bought out, nonprofits invested in) or actively destroyed (through network effects). In the end, both social media site owners and user-activists recognize the same phenomena, but they deal with them in different ways.

The remainder of this chapter uses this framework to explore Wikipedia's history, particularly in terms of the Spanish Fork Labor Strike.

Wikipedia's Discursive-Technical Environment before Spanish Fork

The participants in the Spanish Fork Labor Strike were able to achieve all three associations: they articulated the nascent exploitation of user labor by the owners of Wikipedia, they found pressure points, and they created an alternative. Before exploring how they did so, I want to trace the contours of Wikipedia in early 2001–2002, just prior to the strike, to abstract the discursive and technical ingredients that the participants in the strike drew on to do this work.

First of all, Wikipedia arose as part of a Web-based for-profit corporation during the period when the Interactive Advertising Bureau (IAB) and other groups standardized advertising practices online. In 1996, Wales co-founded a company, Bitter Old Men in Suits (Bomis, Inc.), with Tim Shell and Mike Davis. Bomis began as the dot-com bubble was inflating.[28] Bomis's early ventures included an online directory with content copied from existing open-source material, the Mozilla Directory. This material was modified for a specifically male audience: "Some of its more famous creations were the Bomis Babe Report and the Bomis Babe Ring," along with "adult photo content."[29] Following the newly emerging, standardized advertising model of the 1990s, Bomis sold advertising space around this niche-oriented content.

In addition, Wales wanted to start an online encyclopedia, similarly comprising volunteer-generated material and monetized via advertising. In 2000, Bomis hired Larry Sanger to help create a for-profit, volunteer-driven encyclopedia called the Nupedia. They relied on the dominant model of encyclopedia building: use qualified authors, have a strong editorial policy, and retain copyright. The Nupedia included these elements but was unique in that Sanger and Wales did not hire professional writers; instead, they sought volunteers to write entries. Since Sanger was a newly minted Ph.D. in philosophy, he used his social network in academia to find volunteers. Thus, the Nupedia drew on the Web's overall culture of volunteerism, much as such services as AOL did.[30] However, the project stalled because of an onerous editorial policy that prevented articles from being approved until they were reviewed multiple times (a point I return to below).[31]

Next, Wales was influenced by the Free Software movement of Richard Stallman and the GFDL. Stallman's Free Software movement, which began in the 1980s, is radically opposed to proprietary software and the use of restrictive copyrights to control access to and use of software. Instead of restrictive copyright, Stallman helped create "copyleft" GNU licenses, such as a Free Documentation License. While the original copyright remains with the original author, these licenses allow for others to use, modify, or redistribute a work, so long as any derivative work retains the same license; thus the license becomes reproduced as new work is developed. In this sense, GNU licenses have been called "viral."[32] GNU-style licenses have been fundamental to such software projects as Linux and Emacs, but they have also underpinned open-*content* projects, such as the Mozilla Directory and Project Gutenberg. Because he used Mozilla Directory material for the Bomis directory, Wales was familiar with such projects. His decision to license the Nupedia under a "Nupedia Open Content License," allowing users to copy and modify articles,[33] largely follows Stallman's overall project for free content and software. This engagement with the Free Software ideology contradicts somewhat with the desire to profit from the production of content, but it was necessary to attract free laborers to volunteer for the project. In fact, after negotiation, Stallman came to endorse Wales and Sanger's Nupedia as fulfilling his call for a free online encyclopedia, providing credibility among Free Software activists.[34]

Sanger and Wales used UseModWiki to start another encyclopedia, the Wikipedia. The previous two threads are synthesized here. In 2001, when the centralized, controlled Nupedia project was in danger of floundering because of the slow editorial process, Wales and Sanger set up a wiki-based submission process, open to all, as a way to seed the Nupedia. Moreover, rather than their homegrown Nupedia Open Content License, Wikipedia used Stallman's GFDL, thus gaining more support from the Free Software movement.

Wikis were relatively new phenomena that allowed for the now-famous "anyone can edit" function on the Web. This was a radical departure from the largely read-only model used on much of the Web, not to mention the mass-media model being developed at the time by traditional media firms. Wales and Sanger saw a wiki-based Wikipedia as a side project intended to ultimately speed up their central Nupedia project.[35] To do this, they chose UseModWiki (an opensource software package), set up a domain (Wikipedia.com—note the .com designation, as this was a fateful choice in a Top Level Domain), and relinquished their centralized editorial control to allow for far more distributed creation of encyclopedia entries. Once Wikipedia.com was established, it grew exponentially faster than Nupedia. This rapid growth, coupled with the use of UseModWiki and the GFDL, led to the idea that the Wikipedia could become a global, highly distributed source of information.

However, Wales and Sanger maintained a high degree of centralization that often worked against the distributed structure implied by wikis and the GFDL. Regardless of the use of a wiki and an open editorial policy, Wikipedia remained a dot-com under the control of

Bomis, located in the United States. This centralization was revealed most clearly when Bomis sought to take Wikipedia international; these aspirations were stalled in part because of technological limitations. To enable international participation, Wikipedia's software needed to support non-Latin characters, and the documentation had to be available in multiple languages.[36] However, changes in Wikipedia's software and databases were slow: whereas the editing of Wikipedia articles could, in theory, happen internationally, improvements to the software and access to the database remained more tightly controlled by Bomis, and international editors had to make requests via the mailing list instead of working on the software directly. In addition, UseModWiki was a simple Perl (a server-side scripting language) program, and while Perl is not considered to be a difficult language, few people were capable of modifying UseModWiki to extend its functionality.

Beyond technological limitations, cultural meanings of such concepts as *encyclopedia*, *neutrality*, and *bias* revealed the American-centric aspects of Wikipedia. This can be seen in a debate in the Wikipedia e-mail list: how universal is an encyclopedia? Does it comprise "information" that is independent of language and thus simply has to be translated into various languages? A Swedish Wikipedian named Linus argues for such a scenario: "The information in the encyclopedia should be written in a language-independant [*sic*] manner and then there should be a translation layer on top of that information that adjusts to the wanted language. The biggest problem to solve is the absolute need in this scenario to find a data model that allows article authors to provide their information."[37] In other words, a quantum of neutral, language-independent information could be abstracted from a model article and then machine-translated across all language iterations. In practice, however, this largely meant that such information would have to be pulled from the leading-language version of Wikipedia, which was, of course, the English version. Many Wikipedians working in different languages did not want to simply translate English versions; they wanted to adhere to the complex cultural nuances of their own languages, epistemologies, ontologies, and norms. Edgar Enyedy (who would go on to lead the Spanish Fork) saw the international Wikipedias as independent of the American version, arguing that they should not repeat "the first steps of English Wikipedia…. It's just the idea we want to build our [language version of] Wikipedia from the ground floor, growing as our contributors do, open-minded and open-mindedness of approach from every wikipedia."[38] This issue of translation versus context-specific production would figure into the Spanish Fork.

To sum all this up, Wikipedia's shape before the strike was highly contradictory. Bomis was largely based on the dot-com mass media model of production, and the intended production of the Nupedia reflected this older business model. However, Wikipedia, the side project, was clearly in the nascent Web 2.0 model of more distributed participatory production. Bomis intended to profit from the encyclopedia, but Stallman's influence was felt in terms of the use of a copyleft license and a pressing need to make the encyclopedia free. Wikipedia was meant to be highly distributed across the Internet, but it was also highly centralized, located in servers under Bomis's control of Bomis. Wikipedia was meant to be free, but the meaning of "free" was ambiguous. It was a volunteer-driven project, but it was not a nonprofit (i.e., it was Wikipedia.com, not Wikipedia.org at this point). These intentions culminated in the need for free labor to bring Wikipedia to life. Again, this scenario is rife with contradictions. The use of the GFDL and Stallman's credibility encouraged people to contribute freely to the project.[39] On the other hand, Wales and Sanger's idea to use this freely given work to "seed" the for-profit Nupedia is an early ancestor of the move toward the form of social media capitalism described throughout this book.

The Spanish Fork Labor Strike would emerge within—but not necessarily because of—this contradictory discursive-technological matrix of free labor and capitalism.

The Spanish Fork Labor Strike

In February 2002, the two largest-growing language versions of Wikipedia were the English and Spanish versions.[40] The dominance of the English version was understandable, because, after all, Wikipedia was founded and hosted in the United States. The growth of the Spanish-language version (among others) was an indication that Wales and Sanger's project could expand beyond its American base and go international[41]—unless, that is, international editors went on strike.

This situation is exactly what happened in February 2002. On February 19, Enyedy, a Spanish Wikipedian, announced on the International Wikipedia e-mail list (Intlwiki-L), "Almost every active Spanish wikipedian is on strike till they get a proper proposal [from Jimmy Wales]."[42] The strike was sparked by a brief statement that Sanger had made a few days earlier: "Bomis might well start selling ads on Wikipedia sometime within the next few months."[43] After this mention of advertising, Enyedy replied, "Nobody is going to make even a simple buck placing ads on my work, which is clearly intended for community, moreover, I release my work in terms of free, both word senses, I and want [sic] to remain that way."[44] Wikipedia was a year old at this point, with thousands of volunteer-written articles in multiple languages; clearly, for Enyedy, the idea of this material being surrounded by advertisements was untenable. Shortly afterward, Enyedy and others (including Javier de la Cuena, Juanan Antonio Ruiz Rivas, and Gonis) began to fork Wikipedia by producing a new encyclopedia, *La Enciclopedia Libre Universal en Español* (EL) and hosting it at the University of Seville.

While the possibility of advertising on Wikipedia was the spark that began the strike, it would be oversimplifying to suggest that this issue was the only grievance the strikers had. Enyedy and the others had to associate a wide range of discursive and technical artifacts to justify the fork and enable work to continue on the EL.

In an interview discussing this history, Enyedy calls the centralized structure of Bomis the "American shadow,"[45] which loomed over non-English versions, such as the Spanish-language Wikipedia. The servers were located in the United States, controlled by Bomis. UseModWiki and Perl documentation were written in English, as were the administrative policies (such as the Neutral Point of View policy). The debate over translating English articles into new languages versus rewriting them (discussed above) also made international Wikipedians nervous. These concerns slowed progress in various language iterations of Wikipedia. For example, as Enyedy recalls:

There were significant software issues. The latest software releases and revisions were only installed and running on the
American Wikipedia. The Polish Wikipedia, for example,
could hardly develop at that time due to problems dealing with
special Polish characters. All of the international Wikipedias
were running out-of-date software and because Bomis Inc.
controlled the wiki farm, we couldn't do anything about it.[46]

Indeed, many threads in the Intlwiki-L mail list indicate that
updates from the centralized Wikipedia servers to international versions were slow.[47] Thus, for several international Wikipedians, the
gap between the English version of Wikipedia's software and the international ones pointed to inequality at the organizational and technological levels.

This "American Shadow" was not limited to slow software updating for languages other than English. Enyedy and the strikers also
argued that Sanger exercised too much power over the international
Wikipedias. In their manifesto posted on the EL, titled "Why we
are here and not on Wikipedia,"[48] the strikers noted that "Critics [of
Wikipedia] were censored and Wikipedia administrators used ad hominem attacks" against those critics. Indeed, when Enyedy criticized
the organizational structure of Wikipeda in the Intlwiki-L mail list
(arguing that the project was in danger of having an "'Editor in Chief'
hiding in the shadows"),[49] Sanger responded by calling him a "troll."[50]
Setting aside the issue of ad hominem attacks, the Spanish strikers
were concerned that Sanger was dictating how articles would be written for the Spanish Wikipedia. They were particularly concerned that
Sanger agreed with the call for international Wikipedias to be translations of the English version rather than original works reflecting their
various cultural contexts. As Enyedy recalls:

Larry Sanger acted as a Big Brother. . . . The American
Wikipedia might have seen him as a "facilitator," but we re-
garded Sanger more like an obstacle. At that time he was not
an open-minded person. I have to admit that he brought some
good ideas to us, but the American Wikipedia was too caught

up in the interests of Bomis Inc. I engaged in head-on con-
frontations, open clashes, with Sanger. We were all working
on a basis of collective creation, with peer-to-peer review. It
was an open project, free in both senses. We were all equals,
a horizontal network creating knowledge through individual
effort—this is the most important thing to keep in mind. But
Sanger turned out to be vertically minded. His very status as
a paid employee led him to watch us from above, just waiting
for the right moment to participate in active discussions in the
(mis)belief his words would be more important than ours.[51]

In other words, the ethos of peer-to-peer horizontal collaborative
work, which is a major component of the Free Software ideology ani-
mating wikis and distributed networks, clashed with Bomis's centralized elements. Sanger's status as an employee of Bomis indicated to
the Spanish strikers that any wealth produced by their work may go
to a cofounder of the project. The strikers pointed out this contradictory structure to articulate the budding exploitation of free labor in
Wikipedia.

But the biggest articulation needed to produce a class for itself centered on attention economics and the meaning of "free" in the context
of this nascent form of social media capitalism. When Sanger suggested that Wikipedia be supported by advertisements, Wikipedians
explored advertising's implications in several mailing list threads, both
on Wikipedia and on the new EL. The ensuing debates reveal much
about how the meanings of value production and "monetization" can
be interpreted among different actors. What exactly is a "free" encyclopedia? Does it cost nothing to the reader? Most Wikipedians agreed
on this point. Is it free for anyone to edit? Again, most agreed that
this aspect was essential. Is it free to copy and distribute? The Free
Software ethos of copyleft and Stallman's endorsement of the project meant that Wikipedians assumed this scenario from the beginning. Instead, participants in the Spanish Labor Strike were concerned
about another form of freedom: *is a free encyclopedia free from exploitation of user labor?* For their part, participants in the Spanish Fork
Labor Strike had emphatic responses to questions about the meaning

of a free encyclopedia: it is free to view, free to copy, free to distribute, and, importantly, free from exploitation for profit. As their manifesto puts it, "Culture should not be traded."[52] Ruiz Rivas notes that, once the specter of advertising was raised by Sanger, "we were living the same old story: money, money, money."[53] Enyedy argues, "You cannot appeal to the selfless humanitarian conscience of volunteers while raising money [from advertising] to pay the salaries of Bomis editor-censors."[54] By casting advertising as a form of unfreedom, of exploitation, the strikers were able to heterogeneously engineer volunteer Wikipedia editors as a potentially exploited class of laborers.

Thus, when Sanger announced the possibility of advertising on Wikipedia, many discursive and technical elements individually may not have been alarming to international contributors but when considered altogether (by Enyedy, especially) added up to nascent exploitation. The centralized shape of Wikipedia indicated a potentiality for what Erik Wright has called "organizational exploitation,"[55] where power accrued to Wales and Sanger at the expense of leaders and contributors in other countries. This situation, coupled with the exploitation inherent in monetizing user-generated content via advertisements, was enough to drive the Spanish Wikipedians to strike.

That said, the Spanish Fork Labor Strike was not just a withdrawal of labor. As scholars of open-source software have argued,[56] such projects as Wikipedia that are based on copyleft licenses can be challenged via forking. In this sense, this strike differs from other strikes in that the laborers did not simply stop working; they instead used their labor to establish a competing project. This is a powerful move, akin to strikers who are occupying a factory making the leap from simply striking for a better wage to locking out management altogether and running the factory themselves.[57] Enyedy and the strikers worked to move Spanish Wikipedia content from Bomis's servers to donated server space at the University of Seville. After this occurred, for a period of time in 2002, the amount of content on the Seville-based EL outpaced that of the Spanish Wikipedia.

Ultimately, then, Enyedy and his allies were able to successfully build a class of free laborers for themselves, strike at a key pressure point (Wikipedia's need for free labor, particularly free laborers with non–English language skills), and construct an alternative to lure volunteers away from Wikipedia.

Conclusion: Wikipedia after Spanish Fork

After the strike, Wales and Sanger met many of the strikers' demands. Decrying the antielitism of Wikipedia, Sanger left the project altogether and founded Citizendium, which largely replicated the editorial structure of Nupedia.[58] Sanger's resignation from the project was one of the strikers' key demands. Equally so was their call for Wikipedia to become a nonprofit; in response, Wales stayed with the project and established the Wikimedia Foundation, a nonprofit funded by donations and not by advertising. The issue of translation versus original articles was largely solved by the technological structure of a wiki: if international Wikipedias were to wait for approved English articles to appear to translate, they would never start, because Wikipedia articles are always under revision. Thus, international Wikipedians simply wrote their own articles. The resolution of the Spanish Fork strikers' demands resulted in a truce between the EL and Wikipedia: after a brief period when the EL grew faster than the Spanish Wikipedia, the Spanish Wikipedia ascended to the top ten among all Wikipedias by number of articles, easily eclipsing the EL. Clearly, most Spanish-speaking editors have chosen to (return to) work on Wikipedia.

Thus, the Spanish Fork Labor Strike and its production of a class of users for themselves helped shape Wikipedia into a far less exploitative social media site. This incident represents a departure from the stance of contemporary social media, which present themselves as spaces for *individual* users to shape the system. Wikipedia scholar Andrew Famiglietti sees this distinction: "Wikipedia has not been shaped by the individual desires of radically empowered consumer/producer hybrids, rather it has been shaped by the threat of collective action in the shape of the destructive fork."[59] The class of free laborers involved in Wikipedia were thus able to band together to deny the exploitation of their contributions. "Indeed," notes Nathaniel Tkacz, "not having ads has become a crucial part of Wikipedia's 'free' identity, and since [the Spanish Fork] any talk of ads has always been quickly dismissed."[60]

As a consequence, visitors to the site are not monitored: because it is free of advertising and does not partner with advertising networks or marketing firms, a visitor to Wikipedia does not receive third-party cookies and tracking software on his or her computer.[61] Moreover, in contrast to the constant experimentation in interface design seen on Facebook, Twitter, or Google—experimentation meant to increase the pace at which users produce themselves as consumers[62]—Wikipedia's protocological frame has remained remarkably stable over the past decade. This is due to Wikipedia being divorced from the demands of online marketers; unlike social media capitalist sites, there is no need to tweak the design to increase clickthrough rates on ads, find more optimal placements for them, or discover better measurements of user activities and desires. This stability lends credence to Christian Fuchs's observation that nonprofit media do not need to exploit user data: "If there were no profit motive on internet platforms, then there would be no need to commodify the data and behaviors of internet users."[63] This situation approaches the ideal that Dmytri Kleiner argues for in his *Telekommunist Manifesto*: "Without the need to account for and measure our consumption and production to appease the imposers of capitalist control, workers in a free society may not bother producing exclusively to maximize profit within a 'market economy.' Instead, they may decide to focus their efforts on producing what they want and what their community needs, and are motivated to share the products of their labor out of mutual respect."[64]

This is not to say that Wikipedia is free of surveillance. Rather, surveillance is a key part of Wikipedia, but it is more akin to an ideal of "participatory surveillance,"[65] or a synopticon: everyone is invited to watch pages, watch edits, and watch one another to ensure that Wikipedia's policies are adhered to. The use of wiki software enables editors to trace histories of articles iteration by iteration, and the "Talk" pages allow editors to debate over the production of a neutral encyclopedia. In addition, given that anyone can edit pages—including corporations eager to whitewash articles about themselves or promote themselves—an infrastructure of sousveillance has emerged to combat such edits. For example, Virgil Griffith's WikiScanner[66] cross-references the IP addresses of anonymous editors against a public database of IP address blocks. Thanks to the Domain Name System (DNS), the IP addresses of large organizations, such as corporations, government institutions, political parties, and churches, are publicly available. By using this database, WikiScanner reveals whether anonymous editors are associated with particular organizations. A telling example appeared in 2004, when someone with an Exxon-owned IP address edited the "Exxon Valdez oil spill" page, deleting multiple references to scientific analysis of long-term ecological damage. Instead, the editor inserted "Peer-reveiwed [sic] studies conducted by hundreds of scientists have confirmed that there has been no long-term severe impact to the Prince William Sound ecosystem."[67] On the next edit, the same IP address alleged that the local Native American tribe suffered not because of the death of the local fish population resulting from the spill but rather because of "a series of bad luck [sic] and poor investments."[68] While these edit/IP address couplets do not prove that the edits originated with the management of Exxon, they do show that an employee with access to Exxon's corporate network has made these changes. Surveillance of articles for such potential conflicts of interest and bias is a constitutive part of Wikipedia.

However, despite its nonprofit status, lack of third-party advertising, and distributed sousveillance, Wikipedia still has many centralizing forces. Most apparent is the high level of political-technical knowledge required to edit articles on Wikipedia. The recent case of Timothy Messer-Kruse illustrates this. Messer-Kruse, a historian who specializes in the Haymarket Riot, edited the Wikipedia entry on that topic, only to find his edits reverted immediately. The editors who reverted his work cited Wikipedia's policy against the use of "original research," a category of knowledge production Wikipedia policy defines as "material—such as facts, allegations, and ideas—for which no reliable, published sources exist."[69] This in turn orients us toward Wikipedia's policies on "verifiability"[70] and the "due weight"[71] of sources. To sum up a long story,[72] Messer-Kruse's edits were reverted because his analysis contradicted the bulk of historiography on the Haymarket Riot, and thus his edits violated a cross-section of complex Wikipedia policies and cultural norms. This case illuminates Enyedy's criticism that "Wikipedia has become a huge, hierarchical social network, behind an unreliable knowledge repository."[73] That is, the bulk of Wikipedia's content is administrative rules and Talk pages (spaces where people debate how the articles should be written, which policies Wikipedia should have, and which articles might need to be deleted), not actual articles. This is a crystallization of the extremely complex social structure of the site, a structure that Messer-Kruse confronted (and found to be very different from another complex social structure: academic publishing). It is, in short, a protocological power akin to that of Facebook or Google+: like those sites, Wikipedia is not a space where anything goes. It is a place with many rules, both subtle and explicit, cultural and noopolitical.

This complex of administrative rules, which require long-term engagement and study to understand, creates hierarchies within Wikipedia, and those who better know and implement the rules are drawn closer to the central organizing entity, the Wikimedia Foundation nonprofit located in the United States. Those with sufficient knowledge of the rules can be promoted to a class of privileged users, administrators, and bureaucrats who have more editorial power than the common user. Technically, these people can block IP addresses if they suspect they are associated with repeated vandalism, they can lock editing on pages to prevent edit wars, and they can delete pages outright. Socially, they have a degree of enhanced social capital on Talk pages. Their arguments are often supported by other Wikipedia editors, who look to administrators for guidance on policy issues. At the highest level are the administrators who are associated with the Wikimedia Foundation itself. Called "Stewards," they have technical power over the entire Wikimedia project (from Wikipedia to sister projects, such as Wikibooks and Wikinews). They can alter the wiki templates and ban entire IP blocks (large groups of IP addresses) linked to vandalism. Moreover, this system now has what might be called a "male shadow": studies have shown that the administrators and editors most involved in Wikipedia are males ages eighteen to twenty-five.[74] The exclusion of women, the requirement of understanding complex rules, and the process of drawing the editors most skilled in these rules into the central Wikimedia Foundation quite possibly continue, in modified form, the organizational exploitation that Enyedy and the strikers feared.

Moreover, as Famiglietti[75] demonstrates, Wikipedia's dominance in Google search results and its sheer scale (it is now on hundreds of servers with millions of articles) solidifies its central position as a site for information-seekers. Because Wikipedia entries often appear as the top result for searches, anyone seeking to fork the project to protest Wikimedia Foundation practices and policies would have a far more difficult time than the Spanish strikers did in that first formative year. For example, if feminist editors banded together to create a feminist Wikipedia organized in such a way to challenge Wikipedia's current male-dominated demographics and practices, they would likely have much less impact on Wikipedia's shape than the Spanish Forkers did. Because it would be excluded from high ranking in a Google search, any new project would not approach the power of Wikipedia's network effects; it would not be able to attract a critical mass of free laborers to migrate pages away from Wikipedia and modify them or build new ones. Nor could it compete with Wikipedia's sheer scale. At best, such a protest would achieve the stature of other Wikipedia forks, such as Conservapedia, an online encyclopedia created by conservatives to combat perceived liberal bias within Wikipedia. Wikipedia is now firmly established:

As projects persist over time and space, they garner new participants, make and fix mistakes, develop and argue over policies, secure regular funders, become embroiled in media scandals, celebrate milestones, and generally extend outwards, becoming more real. Their forensic reality is amplified; their boundaries grow, shift, and are difficult to locate. The task of generating equivalences [that is, the task of producing a competing fork] becomes more difficult.[76]

To be effective, another strike against Wikipedia would have to be massive.

Thus, Wikipedia now appears to be stabilized, but, as Dugdale might argue,[77] this stability hides uncertainties and tensions. The contradictory tensions that erupted in the Spanish Fork Labor Strike still oscillate within Wikipedia. But Wikipedia's history does provide insights for a more egalitarian, democratic Web. The Spanish Fork Labor Strikers' successful engineering of a class for itself is in many ways now built into the software and structure of Wikipedia.

In the final chapter, I consider current efforts that in many ways follow and extend Wikipedia's lead by heterogeneously engineering nonexploitative and free social media software and cultural production and, along the way, help continue to construct a user class for itself organized against the commodification of daily life happening on such sites as Facebook, Google, and Twitter.

6

A Manifesto for Socialized Media

> If we had a real intellectually-defensible taxonomy of
> services, we would recognise that a number of the services
> which are currently highly centralised, and which count
> for a lot of the surveillance built in to the society that we
> are moving towards, are services which do not require
> centralisation in order to be technologically deliverable.
> They are really the Web repackaged.
>
> —EBEN MOGLEN, in Glyn Moody and Eben Moglen,
> "Interview: Eben Moglen—Freedom vs. the Cloud Log"

This book has been about the heterogeneous engineering of social
media software as it has been produced in capitalism. It must
also be about resistance to the inequalities and reductions built into
that system. It must be about potential ways to dissociate social media capitalism. It must, therefore, be about the *reverse* engineering
of social media software. To this end, this chapter attempts to synthesize two things: a design for an ideal social media system and the
specific, material efforts to build such a system. In much the manner
described in Chapter 3, it thus proposes an abstract architecture and
then considers the concrete ways in which that architecture might be
implemented.

Returning to the metaphor of reverse engineering, we must first
ask: why bother? If social media are exploitative and reductive, why
even try to start with their current designs to build another system?
As Kathryn Ingle notes in her book *Reverse Engineering*, when we
confront a system that we want to replace, "every time you have a
better idea you will consider all the positive design aspects before
condemning an entire product."[1] That is, we do not simply throw
away existing technologies; we critically assess them to discover properties we value in them. Turning to another context, for all of Karl

Marx and Friedrich Engels's rage about capitalism, the first parts of *Communist Manifesto* reveal the revolutionary power of capitalism to topple a previous oppressive regime, feudal monarchy. The *Manifesto* is not atavistic; it does not long for a lost Edenic past that has been trampled by the bourgeoisie. Similarly, in terms of technological change, it is pointless to long for a time before social media, some lost period where John Perry Barlow's *Declaration of the Independence of Cyberspace* is an empirically accurate governing document. We must work with the technology we have. In fact, this is a lesson of Occupy Wall Street and the Arab Spring. Activists in these movements have shown us how to use a wide range of communications tools—including hegemonic social media—to resist the very system that gave rise to sites built on the exploitation of user labor. It is a gross and insulting reduction to brand the Tunisian and Egyptian revolutions as "Twitter Revolutions," but certainly all these social movements have drawn on Twitter, Facebook, and YouTube. Beyond social movements, the sheer popularity of social media, their pleasures, and their challenges to mass media cannot be ignored. Indeed, any reverse engineering of social media that has the goal of building a better media system cannot ignore these elements.

So, the first part of this chapter is a manifesto for what I call "socialized media." Working from the assumption that democracy requires debate and collective production of knowledge, and working from the observation that mass media offer neither of these but social media have demonstrated the potential to do so, I argue for a system that involves true two-way communication, decentralization, free and open-source software, and encryption. I also argue that socialized media will require a radical pedagogy at their interfaces that can guide users through the layers of abstraction that software comprises and teach them how to modify it to meet their needs. Similarly, I argue for the dissolution of the producer/user distinction, suggesting instead that socialized media rely on collaboration between users with and without technical skill to co-construct it. In opposition to the increasing surveillance within social media, I call for a culture of fluid identities, for an antiarchival system, and for the eradication of intellectual property. Finally, to oppose the obsession with the new within social media and the culture of consumption outside it, I argue that socialized media must be platform-independent (they must run on any machine) and, moreover, that they should run on free hardware—free as in free beer. I propose that there is one benefit to the irrational overproduction of computers, smartphones, routers, and networking equipment—evidenced by the glut of e-waste dumped on the Global South and necessitating the destructive labor practices of mining and factories that produce gadgets—and that benefit is this: the world has plenty of computer equipment on which anyone who wishes to could install socialized media software.

Returning to Chapter 3, which explored the relationship between architecture and implementation, I hope it is clear that my idealized design is more speculation than specification. This speculation arises from the critical analysis of existing social media as well as its historical context of software engineering and global capitalism. Since I am playing the role of design speculator, I must prepare myself for the inevitable resistance from implementers to the architecture I propose. Here I am thinking of Fred Moody's *I Sing the Body Electronic*, an instructive look at the gap between design and implementation. Moody spent a year in the early 1990s with a Microsoft team that was building a children's multimedia encyclopedia. His book reads as a catalog of knowledge-worker angst and irrational exuberance. Somewhere in this bipolar motion, there was a contradiction between designers (who dream up features) and developers (who have to code them). Nearly every page introduces a new feature from a designer, to which the coders respond, "That can't be done." This contradiction threatened to dissociate the project.[2] Similarly, in many ways, my socialized media system is a dream that would probably make a coder shudder!

Nevertheless, in the second part of this chapter, I explore current, existing alternatives to dominant social media as potential implementations of the design I propose here. Unlike the children's encyclopedia in Moody's book, I hope that the gap between the ideal design and material implementation, and the tensions that come with it, are generative. Moreover, those building social media alternatives themselves often hold political and activist ideals that they struggle to embody in code. The gap between an ideal system and the material circumstances of the world always exists; the question is this: what do we do to synthesize the material and ideal? I hope my all-too-brief survey of existing social media alternatives illuminates ways in which activists are attempting to bridge the gap.

Idealizing a Socialized Media System

First, to my design speculation: I offer a feature set for socialized media.

Equal capacity to receive and transmit. This is probably the most obvious required feature, one that Hans Enzensberger argued for in 1970 in laying out a Marxist theory of media: we need to exploit the fact that such devices as networked computers easily transmit and receive.[3] In many ways, this is the dream of World Wide Web inventor Tim Berners-Lee: the Read/Write Web, where every browser has both the capacity to render pages (i.e., receive) and alter them (i.e., transmit).[4] Berners-Lee wanted open creation and distribution in the 1990s, but this largely did not occur. Instead, advocates of Web 2.0 argue that the Read/Write Web that Berners-Lee wanted arose during the early 2000s in social media; Web pages now include such interactive features as commenting, uploading, and the like. As should be clear, I have argued throughout this book that, while it is compelling, this two-way transmission has been severely curtailed within social media via surveillance, centralization, and protocological control. Despite their exuberance, social media cheerleaders proclaiming the democratization of media are missing these inequalities, even as they are right to point out that social media are in many ways the break from mass media that Berners-Lee envisioned.

Beyond the structural inequities I have explored in previous chapters, the ideal of two-way communication on the Internet is also being undermined in a manner that Enzensberger would readily recognize: in the current broadband network architecture, there is often a large gap between upload speeds and download speeds. Download speeds are faster, and they are advertised as indicators of which network is better. For anyone seeking to watch streaming video or download music, a high download speed is no doubt helpful. But in a true two-way architecture, the networked computer can *upload* data just as fast as it can receive it. As in the history of radio, where two-way radio communication was regulated out of existence in favor of the top-down commercial model we enjoy today, this gap between upload and download speeds should trouble us, and no such gap should exist in socialized media.[5] My upload speed should be the same as your download speed as we connect in two-way communication.

Thus, socialized media must be at its core designed to provide equal transmission and reception capabilities, and its software must be built so that inequalities between these two can be avoided or worked around.

A decentralized architecture. Two-way communication is no guarantor of equality; we can easily imagine a scenario where multiple two-way devices are aggregated in a central location, creating network effects and gaining power over the traffic that flows through that location. This is what we see with the massive server farms of Facebook, Google, and Amazon. Moreover, as we have seen in the shift from dial-up to broadband Internet service providers (ISPs), the very connections we use to get online and engage in two-way communication are getting centralized as well. ISPs now represent a disconcerting accumulation of power, evidenced by the increasing use of Deep Packet Inspection (DPI) and packet discrimination.[6] In fact, the recent political and lobbying battles over Network Neutrality have revealed two great poles of network centralization: on the one side are Google, Facebook, Apple, and Amazon, whose server farms belie the idea that any garage-based individual can compete with them in terms of raw computing power, low latency, and storage; on the other side are ISPs, such as Comcast, and network hardware manufacturers, such as Cisco, that seek to shape the network to better fuse dominance in intellectual property distribution (Comcast does have a desire to have us watching NBC instead of homemade YouTube videos) with a network that favors mass media and metered communication over user-generated or pirated content and unmetered, free communication.

Thus, socialized media must be decentralized and, more importantly, decentralizing, always threatening to dissociate nodes of power, such as server farms and delimited access points. This is the power of peer-to-peer software and networked computing: two-way communication and computation are aided by mobile, spontaneous networks capable of circumventing blockages and relaying information along shifting and changing paths. This resiliency was, in fact, the mythical impulse of packet-switching networks, such as the ARPANET, a structure severely curtailed in the centralized network we now use.

I am mindful of Langdon Winner's critique that "'decentralization' is one of the foggiest, most often abused concepts in political language."[7] As the adage goes, it is easier said than done:

> Dreams of instant liberation from centralized social control have accompanied virtually every important new technological system introduced during the past century and a half. The emancipation proposed by decentralist philosophers as a deliberate goal requiring long, arduous social struggle has been upheld by technological optimists as a condition to be realized simply by adopting a new gadget.[8]

Decentralization is not simple, because it cannot be achieved by buying a new gadget or, as we saw in Chapter 4, by uncritically producing distributed protocols and building decentralized networks. If pockets of power aggregate in the form of an elite within an organization, that power can be hidden behind the discourse of equality and decentralization—perhaps more effectively so than in cases where power is more bald-faced. Moreover, as Alexander Galloway has convincingly argued, social power and control do not disappear after decentralization.[9] In decentralization, power accrues to those who shape protocols, because protocols regulate connections between distributed nodes. For example, Internet governance is putatively open to all, but it is largely the purview of technocrats who can read and write the technical code that composes the protocols. To be part of the network, one must use these protocols, and thus one must be subject to whatever modulatory power is embodied in them.

This leads to the need for an *open-source architecture*. Simply put, the software, protocols, and power structures of socialized media must be open to inspection, alteration, extension, and distribution. To use an overused and fraught term, there must be transparency in how these systems work, which policies and technical limitations constrain them, their iterations through time, and the arguments made by those who propose or oppose changes in their design. This transparency must also extend to the governance and political economy of these systems, where fluid mixtures of democratic and expert decision making must occur. Thus, the software, hardware, and culture of socialized media must be open to modification.

Again, this requires a peer-to-peer, distributed structure as opposed to the client-server architecture. After all, the servers in the Google server farm are likely running open-source software and languages such as Linux, Apache, MySQL, PHP, and Perl. However, no matter how open the source code is, the server farm is largely inscrutable to a Web browser; even with the "View Source" option in the browser, one cannot see the contents of any server. In contrast, a peer-to-peer distributed system would run on peer computers, allowing the owner of the computer to inspect and alter the code as well as have a reference point in debates about how the system be administered.

But again, open source is not necessarily inclusive. For example, reading and writing code is a specialized skill, and thus even with open-source software, the barrier to participation is high. A famous—and disappointing—acronym in Internet forums dedicated to support for open-access software is RTFM: Read the Fucking Manual. This happens when a "n00b" open-source user asks a question that a "1337" open-source hacker has (to be fair) answered many times in other threads. This micromoment of condescension reflects a larger elitism that can frustrate computer users and turn them back to the world of proprietary software, a world in which glossy abstractions, such as lowest-common-denominator user interfaces, hide the complexities of the closed software beneath. In a bizarre inversion (and similar to Winner's point about the term "decentralization"), in much of the software engineering literature, glossy abstractions that do not reveal their depths are "transparent," even though such transparency can hide Digital Rights Management (DRM) limitations and surveillance that subtly limit the uses for the Universal Machine.[10] Despite this, ease of use and "the user experience" is so emphasized in computation that it would be foolish to ignore it as an architectural goal, even if it means that TFM must be well-written and accessible and that n00bs get more respect in online forums as they ask questions about the software installed on their computers.

Beyond being nice to n00bs and producing good documentation, I want to propose something a bit more radical: the architecture of socialized media also needs a *radical democratic pedagogy at the interface.* One of the advantages of proprietary software has been that it has a built-in pedagogy; Frederick Brooks (of IBM and *The Mythical Man-Month* fame) expressed a sense of wonder in seeing the pedagogy of graphical interfaces, noting that PCs with graphical user interfaces (GUIs) invite us to simply play with them to learn how they work.[11] Indeed, the radical break between the PC and mainframes centered on the elimination of batch processing and obeisance to what Paul E. Ceruzzi has aptly called "the priesthood" of computing[12] in favor of the "Californian Ideology" of libertarian, individual power over the machine.[13]

However, the pedagogy of the PC is deceptive; the shift from batch to personal computing has fit well with ideologies of neoliberal individualism that elide both the proprietary software running underneath the surface and the material, social, and ecological impact of the craze for individual gadgets.[14] This deception and atomization have only extended as networked computation shifts from PCs to smartphones. In contrast to this, the interfaces of socialized media must provide pedagogies that teach users two things: first, how their machines and software are linked to one another, how that linking offers greater power than any machine/user working alone, and how that linking includes a global network of material and immaterial flows. The surface of socialized media software and hardware cannot hide its reliance on energy sources, the labor of its construction, the networks of its distribution, or the conditions of its disposal. Nor can it hide the machine/user's dependence on others to achieve social goals. Second, this graphical pedagogy should guide the user past the surface of the software through the depths of the machine, through the (legal, political, economic, software) code that animates it, and the ways in which that code can be altered for social ends.

All this said, clearly the interface must draw on the conventions of past interfaces to provide new users with (to use an abused term) intuitive control over the software. Ease of use and aesthetics cannot be discounted, and these conventions should be selectively used to teach new users how their machines and software work and how they can become involved in shaping them to meet their needs. Moreover, such interfaces and software must be well documented (in as many languages as possible!). Documentation, which can explain how the object works and, moreover, what politics it contains, is as important as the software itself. The production of such documentation has been eased significantly with the advent of wikis.

This radical interface pedagogy would aid in creating *a collaborative collective of users with and without technical skill.* Steve Woolgar's ethnographic analysis of a computer company reveals the practice of "configuring the user" as an "outsider," contrasted with the company's employees, who are insiders in the know about the technology they produce.[15] To comprehend the user, the company makes the user a flattened and homogenized Other. This practice displaces the inherent heterogeneity in the wide range of people who might come into contact with the technological artifact and use it in creative ways. This is a boundary issue, symbolized by holographic stickers on the seams of machines that say, "If this seal is broken, the warranty is void." We might even say that this is deuniversalizing the Universal Machine, making it delimited in ways that enable the company to maintain power over the technology it produces. As Andrew Famiglietti argues:

In contrast to the cyborg individual, who dominates his or her machine and is free, the user is dominated by the machine, and thus enslaved. This othering of those who do not display technical mastery may serve to blind communities shaped by the cyborg individualist ideal to the possibilities of building meaningful connections with those outside of a narrow range of technical proficiency.[16]

In contrast to this, socialized media software must blur and collapse inside and outside, user and producer, in ways that are wholly different from the hegemonic "produser" idea now used by current social media companies.[17] Coders and designers, so-called digital natives and those afraid of these newfangled machines, modernists and postmodernists,[18] should all be allowed to modify the system as they see fit. Returning to Enzensberger, this is "democratic manipulation,"[19] the democratic production of meanings in all forms, from code to comments, coming from people of myriad backgrounds and abilities. It is a relationship not between producers and consumers but between citizens and citizens, working to build a media system worthy of democracy, educating and debating one another about the shape of their collectively produced media system.

Obviously, this means that socialized media need *copyleft, not copyright.* The holographic sticker warning would-be tinkerers against voiding their warranties is a tiny part of a vast intellectual property (IP) system organized against the copying, manipulating, and distributing of a whole host of productive and cultural artifacts. This is an ethos of exclusion, a necessary part of a logic of private property–based exploitation.

This system of artificial restraints is probably the biggest counterweight to any true two-way peer-to-peer system, because such systems have allowed for sharing and distributing digital goods. IP-holders are demanding that states do something about such piracy, and states have responded with increased surveillance of online activities. However, rather than simply proclaim that such surveillance is to catch movie uploaders on behalf of Disney and DreamWorks, states use the enemy du jour to justify their work. Consider the Orwellian shift in enemies and rationales: Eurasia's Terrorism is the enemy one week, and Eastasia's China (or specifically, "Chinese patriotic hackers") is the enemy the next. One week, cyberwarriors will blow up dams and disrupt the electrical grid; the next week they will hack into "the network" and steal government secrets. Looking at the arguments of cyberwar drumbeaters over time, it seems as though little memory of one claimed enemy or attack vector is retained from week to week.[20] The only consistencies in rhetorics of cyberwar and cyberattacks involve the repeated assertion that we will be attacked and that—interestingly enough—we must protect exclusive intellectual property from pirates at all costs.[21] Seemingly moving alongside terrorists and patriotic hackers, online pirates are putatively destroying whole economies by downloading movies and music. So long as IP-holders continue to hold onto a vision of property as exclusion, they will pour money into lobbying; writing legislation, such as the Cyber Intelligence Sharing and Protection Act (CISPA); and hammering out global trade agreements, such as the Anti-counterfeiting Trade Agreement (ACTA), while defense agencies, such as the U.S. Department of Homeland Security (which has seized many Internet domain names to prevent piracy) will continue to push for expanded capacities to monitor online activities, claiming to be fighting a "cyberwar" (whatever that means) while actually prosecuting music uploaders.

All of this is not to downplay the danger of such ubiquitous surveillance. This is especially true now that Edward Snowden has leaked internal details of the National Security Agency's (NSA's) massive monitoring systems. The centralized social media (such as Facebook and Google), ISPs, and cell phone providers (such as Verizon) make easy targets for warrants, National Security Letters, or illegal wiretaps. The ease of copying and storing Internet traffic (as the National Security Agency is doing at Internet Exchange points[22] and in a massive server farm in Utah)[23] means that there is less and less need for targeted surveillance of suspects; instead, all data everywhere will be simply stored for ex post facto analysis. We can all be watched, and we can all tell ourselves we have nothing to hide.[24]

But this gets away from the central point about IP. In contrast to a vision of property as the right to exclude (and thus as something that requires a strong state to monitor and prosecute citizens who violate property rights), socialized media must be based on copyleft licensing that defends the right to distribute and the right of collectives to productive property.[25] This brings creation closer to the ideal of production for use and not for profit, because it reimagines humans as "incentivized" not by greed but rather by creativity and the chance to improve their lives and their communities. It would also severely reduce the need for a state that polices property violations; what is there to police when the tools of cultural and digital production are freely available to all? The old mass media power has rested on the creation, alienation, protection, and sale of intellectual property as its economic model, and thus it constantly captures regulatory power to increase copyright lifespans and monitor networks for pirated exchanges. This system is in need of sloughing off in favor of the creative commons.

The war on piracy, a corollary to the larger, growing regime of surveillance within social media, brings me to another point: the socialized media need *encryption*. In some ways, this is already a feature of contemporary social media. The Firesheep controversy, where a programmer built a Firefox extension that could capture plaintext passwords sent across unsecured wireless networks, prompted Facebook and Twitter to offer HTTPS connections as well as the Electronic Frontier Foundation's HTTPS Everywhere browser plugin (which forces sites to use encrypted connections wherever possible). Encryption has also been a long-standing feature of e-commerce sites and can be used in e-mail clients. Peer-to-peer distributed socialized media will expand on these practices, using encryption across the network. Each user's machine will encrypt the data stored on it as well as the traffic it relays across the network.

This need for encryption may seem to contradict my call for an open-source architecture, but I follow Christian Fuchs in his distinction between the privacy of ownership and power versus the privacy of citizens and workers. As he argues in his analysis of Facebook, we cannot allow for potential and actual domination and power to be opaque. Instead,

> economic privacy should be posited as undesirable in those cases, in which it protects the rich and capital from public accountability, but as desirable, in which it tries to protect citizens, workers, and consumers from corporate surveillance. Public surveillance of the income of the rich and of companies as well as public mechanisms that make their wealth transparent are desirable for making wealth and income gaps in capitalism visible. Such an approach includes privacy pro-

tection from corporate surveillance. In a socialist conception of privacy, the existing privacy values have to be reversed.[26]

Thus, the coupling of transparency in the open-source architecture with encryption (as well as cultures of fluid identity and new conceptions of the archive, points I explore below) will reverse the power flows that contemporary social media currently comprise.

Socialized media software must also be *platform independent.* It must run anywhere, whether on rebuilt computers or jailbroken phones. In fact, socialized media not only should be platform independent and open source but also should run *on free hardware.* This means free as in free beer. With the glut of electronic waste being dumped on the Global South—waste comprising in large part computers and cell phones—and with garage shelves and drawers in the Global North stuffed with used (but "obsolete") computers, a huge supply of hardware is available for a socialized media system.[27] As Gavin Mueller has noted, the argument against free material commodities has hinged on the economic theory of rivalry: if I own a computer, I may exclude you from using it. In contrast, of course, IP-holders are anxious because of digital artifacts' nonrivalrous properties: if I have a digital copy of a song, there is no reason you cannot also have a copy. Digitization of IP provides us with a new perspective on the sheer number of commodities produced within capitalism. If there is one benefit of the manic production of new consumer goods, it is that we now know how to produce such material in mass quantities and that we each could have the ones we need or want. Because of mass production, a mass of nondigital artifacts mirrors the mass of digital ones. As Mueller writes:

> While it may be true in a limited metaphysical sense that [objects like] shoes are rivalrous, in the context of the actual world (from which all analysis should proceed), we know that vast quantities of shoes are produced. . . . Even if other people bought shoes, it wouldn't prevent me from getting the exact same, or a nearly identical, type of shoe. So this kind of overproduction of mass-produced goods, combined with the

widespread inability for many people to pay for these goods, renders the point about rivalrous goods moot.[28]

The inequality between haves and have-nots vis-à-vis such objects as shoes or computers is only amplified when those who can consume them dispose of them in dumps located, oddly enough, in areas of the world where poor people live.[29] This inequality is only amplified further when it turns out that much of what is being dumped *still works*. Although the purchase of a computer or smartphone is an individual act, its consequences have been socialized; it is time to socialize production of these materials. Any socialized media system would use these machines first to build networks rather than dump them into putative "recycling" centers that send them to the Global South only to be burned so that children can pick them apart to retrieve bits of precious metal while the by-products of such "recycling" poison air, water, and lungs.

Socialized media would thus hinge on reducing the sociopathic emphasis on the new within social media. A corollary to (or perhaps a determinant of) streams of the new appearing on Twitter and Facebook is the lust for new gadgets. As Richard Maxwell and Toby Miller argue, "Rapid but planned cycles of innovation and obsolescence accelerate the production of electronic hardware and the accumulation of obsolete media, which are transformed overnight into junk. Today's digital devices are made to break or become uncool in cycles of twelve months or counting down (check your warranty)."[30] iPad 3Gs beget 4Gs beget Minis, notebooks beget netbooks, flip phones beget smartphones, and all the while miners must work in conflict zones to extract the increasingly rare materials they contain, workers must build these new gadgets in brutal factories, consumers must "keep up" by buying the latest versions, and dumps continue to fill. However, despite Moore's Law or Kurzweilian techno-religions built on exponential change, computer architecture has largely been stable since the days of the Von Neumann Architecture. And, despite Maxwell and Miller's point about electronics being built to break, many of these older pieces can be resuscitated. Thus, older computers with new operating systems and hardware installed on them will network together and run quite well.[31] The goal of socialized media is to reengineer our relationship to the new and to think about durability, longevity, and the production of hardware for use, not for short-term profit and long-term ecological destruction.

Stepping away from hardware, we shift back to culture. We want to eradicate the reduction of individuals to consumerbots operating under the gaze of surveillance systems, and yet we value social media's public performances of identity and affect.[32] To navigate this, we will also need a culture of *fluid identities, with anonymity, pseudonymity, identity shifts, and play*. The dominance of Facebook has led to a culture of real-world identities in social media. If this model is coupled with individualized technologies (such as smartphones and tablets) and network analysis, it becomes far easier to monitor, typify, and reduce users to mere consumption machines. Instead, socialized media would welcome, not discourage, users' abilities to play with their identities: categorizations, metadata, and XML schema would be either completely open-ended or simply ignored in favor of completely user-defined systems. Socialized media also cannot be templated media; the templates of contemporary social media reflect the underlying standardization of user data and the disciplining of their activities. Instead, socialized media would be just as open-ended and pliant as HTML and CSS. Indeed, we might see more strange and dangerous "pimped" profiles, but that is the price of a collective sense of privacy and freedom.

Similarly, socialized media will need an *antiarchival system*. In current social media, we do not have real-time surveillance so much as what might be called ex post facto archival surveillance, where "facts" about individuals are produced by abstracting items from databases and remediating them in such settings as courtrooms and marketing algorithms.[33] Even with a culture of fluid identities, bits of information gathered through surveillance can be articulated into a real-world identity or, more properly, the disciplined identity of what Gilles Deleuze famously called the 'dividual—a subject divided into easily controlled quantities, categories, and attributes.[34] We must imagine an archive of data in which every item is linked to something that contradicts it, that undermines its potential operation as a fact in the world. In other words, we cannot have archives that could be used to construct and discipline an interactive consumer or raided by states seeking damning evidence on dissidents. We need archives that can produce only material that undermines any attempt by abusers of power to construct particular (and thus condemnable, controllable) images of individuals. Following Deleuze, we might call this "anti-'dividuation."[35] The objects stored in the database must appear in ironic, shifting, unreliable, and whimsical forms. Perhaps the model for this would be the idiosyncratic, premodern Wunderkammern instead of rationalized archives. Or, perhaps this would involve democratic ontologies, so-called folksonomies, but ones that deny rationalization by endlessly deferring the production of facts.

This would also alter our relationship to time in social media; rather than the new and immediate being a moment of constantly updating our statuses, perhaps random elements from the past would crop up and confront us unexpectedly and out of context. Perhaps all data—and, importantly, metadata—should be made for forgetting. This is less about a social stream of new updates and more about a different relationship to memory and to one another. We can imagine metadata that do not persist but rather die after a short period of time—metadata with Time to Live. To be sure, it is very possible that computers cannot do this, that they are too rooted in Western rationality and strict relationships to enable such a system. But given the power that accrues to anyone who can build archives of data out of network flows, some antiarchivism in socialized media is much needed.

With this (very likely too ideal) design in mind, it is time to take off the designer hat and see how actual implementations of these ideals are being developed.

Materializing the Ideal: Existing Social Media Alternatives

The dominance of Facebook, Google, Twitter, ISPs, and surveillance agencies is not going unchallenged. A wide range of social media alternatives is being developed or has, in fact, existed for years. Examples include, but are not limited to, Diaspora, GNU Social, Freedombox,

TalkOpen, Tor, Facebook Resistance, meshnets, Yacy, Ixquick, Crabgrass, Creative Commons, Move Commons, Riseup, Zurker, and
Lorea. What follows is an all-too-brief survey of these systems, using
the design speculation above as a framework for discussing them.
Along the way, I want to consider these systems as artifacts being heterogeneously engineered. How do people associate technology and
ideas into these systems? How do they relate to the ideal? To the aesthetic and code conventions of the monopolies? To one another? To
activism and social justice?

Perhaps the most common technical and discursive association
being developed by alternative social media is a decentralized system.
"Peer to peer," "federated," "distributed," and "decentralized" are powerful rhetorical terms that can help attract the necessary activist coders and users needed to implement a socialized media system. Serious
projects dedicated to building distributed social networking sites on a
peer-to-peer architecture are underway. Lorea, for example, is a project to build a federated media system inspired by Deleuze and Félix
Guattari's concept of the rhizome. As Lorea's About page puts it:

> In simple terms, this is the idea that we do not need verti-
> cal or horizontal structures that require the adoption of one
> ideology by all comrades. We can bring together all comrades
> and parties, all of whom are acting and desiring subjects, and
> by bringing them together create a whole that is greater than
> the sum of its separate parts. Within N-1/Lorea,[36] each will
> learn to add variety and heterogeneity without bowing to any
> unique or unequivocal truth.[37]

Another is Crabgrass,[38] which takes a different tack on social
media, using the group rather than the individual as its main organizing principle.[39] Thus, its two-way communication principle is
different from the individualized (or 'dividuated) forms in dominant
social media. As Elijah Sparrow notes, Crabgrass's builders are (to use
a software-development term) eating their own dog food: they use
Crabgrass to organize their group-to-group political activities.[40] Both

Lorea and Crabgrass are projects to build and extend free software that can be installed on multiple servers. Lorea, for example, is built out of Elgg, an open-source suite of social networking tools.

Finally, Diaspora is a more mainstream decentralized social media system. In that system, it is possible to set up one's own server (called a "pod" in Diaspora parlance) and run Diaspora on it, but many people are using a central JoinDiaspora.com server. One could imagine Diaspora centralizing via network effects onto this one server, but the fact that it is open source might make it more akin to WordPress, the open-source blogging software that can be installed on any server—that is, if users believe that the administrators of JoinDiaspora are taking the site in an exploitative direction, they can flee to their own servers, much as the Spanish Wikipedians did (see Chapter 5). Unlike Lorea and Crabgrass, however, Diaspora is seeking venture capital, raising the specter of commodification of user data. It is not clear how venture capitalists would profit from investment in the company, which is a nonprofit dedicated to building open-source software.[41] Given that its slogan is "you own your data," then the "data locker" business model of creating a marketplace where users can sell their data is one possibility.[42] However, because it uses a copyleft license, at the very least the Diaspora code can be appropriated for nonprofit or collective-profit use.

Even a decentralized peer-to-peer system is powerless when access to the network is monitored, as is the case within many nations around the world, or eliminated altogether, as happened when the Egyptian government shut down the country's five ISPs in 2011. To combat this, activists are building anonymizing systems and meshnets. The most notable of the former is Tor, an onion-routing system that provides an encrypted connection to a network of nodes. Like many Internet technologies, it was developed by the U.S. military but is now available for civilian use. The "onion" in onion routing refers to the multiple layers of encryption placed around packets of information; each layer is peeled back by the various routers the packet traverses, until the unencrypted packet reaches its final destination. If the final destination also uses encryption (i.e., TLS or HTTPS), the data are encrypted for the whole circuit. Moreover, Tor operates as a proxy, obfuscating a user's IP address, thus allowing users a way around network firewalls and DNS blockages.

Meshnets are ad hoc, self-configuring, self-healing networks of computers using wireless connections. As opposed to the ISP model, where a computer connects to an ISP and then is connected to the broader Internet to upload and download data, each computer in a meshnet relays others' data as well as uploads and downloads its own. This distributes control over access to the Internet. As Julian Dibbell puts it:

> If you want a better sense of what that means, consider how things might have happened on January 28 if Egypt's citizens communicated not through a few ISPs but by way of mesh networks. At the very least, it would have taken a lot more than five phone calls to shut that network down. Because each user of a mesh network owns and controls his or her own small piece of the network infrastructure, it might have taken as many phone calls as there were users—and much more persuading, for most of those users, than the ISPs' executives needed.[43]

Meshnets do face technical limitations: for example, "the available [Media Access Controls] and routing protocols are not scalable, [and] throughput drops significantly as the number of nodes or hops in [wireless mesh networks] increases."[44] And there is the sheer effort required to enroll other users and sundry technologies into a successful association to build such a network. However, these systems promise to mitigate against centralized ISP control.

Of course, returning to Winner's warning,[45] decentralization is not easy. In terms of the technical infrastructure of these sites, immediate problems of implementation arise: how do members find one another?[46] Client-server systems easily allow contacts to find one another. Peer-to-peer systems may have to rely on a central directory server, which brings us back to the centralization problem. Or, they may use the peer-to-peer technology Distributed Hash Tables (developed to allow file-sharing services to find files across ad hoc networks); however, this is a difficult system to implement.[47] One answer to finding friends' digitized avatars may rest on decentralized search. An ambitious project here is YaCy, a peer-to-peer search engine that can be installed on any computer using Java.

With these problems, it seems to be easier to dream of decentralized architectures than it is to build them, and many social media alternatives have, in fact, retained the client-server architecture. For example, Lorea remains on the central servers located at n-1.cc. TalkOpen, a short-lived alternative to Twitter, also existed on a central server. In the case of TalkOpen, instead of promising decentralization, the site relied on an extremely simple and unequivocal privacy policy[48] to attract a membership that was highly sympathetic to the hacker collective Anonymous. This policy was meant as a counterweight to the key problem of client-server: the ability for a state to demand or seize data from the service. In a sense, this policy was a reduction of central power not by distributing it but rather by denying it in the first place. Such search engines as Ixquick, Startpage, and DuckDuckGo are similar in that they do not log user IP addresses. This active denial of basic archiving practices is perhaps the closest practice of my argument for an antiarchival system. However, in all these cases, one can imagine that state power can overwhelm these sites; TalkOpen's or Lorea's administrators might be compelled to allow a state to surreptitiously monitor user activity (in the U.S. context, think of the anonymous power of the National Security Letter), and these sites might be compelled by law to keep logs for security reasons.[49]

Another alternative, Zurker, is centralized in the server-client architecture but attempts to distribute power by providing ownership shares to members. Members can receive shares by inviting others to join or by purchasing them. Zurker's accounts are open, allowing members to examine incoming and outgoing expenditures.[50] Clearly, then, this is akin to open-access transparency that has been a rallying cry of media activists. However, as a shareholding system, one can imagine that the democratic participation in Zurker is akin to democracy after *Citizens United*: if a dollar is a vote, then those with the most dollars/shares get the most votes. Moreover, a shareholding system may imply the classic venture-capitalist practice of building a site and then selling it or putting it up for an IPO on global stock markets. In either of those cases, since the predominant economic model of social media is monetization via reducing users to marketing segments, Zurker may simply end up being the social media equivalent of multilevel marketing.

However, the biggest hurdle to the heterogeneous engineering of an alternative social media system—decentralized or not—is the network effects of the monopolies. People use Facebook, Google, and Twitter because everyone else does. Returning to the art of sumoto. iki (described in the introduction to this book), hegemonic social media's "ghostly frames" haunt any efforts to reverse engineer social media and build alternatives. Because everyone else uses them, there is a great deal of cultural knowledge of how these systems work and which features they offer. And because a user base is a sine qua non for social media, those building alternatives often note that they must imitate the monopolies to lure users away. This challenge gets at the disparity between technologically adept users, who can install open-source software, manage servers, and encrypt connections, and those without such skills. Much of social media's popularity has arisen from its simplicity; the simplicity of the Google home page is a classic example, and (as I explored in Chapter 3), Facebook's clean aesthetic helped it overcome its rival Myspace. This simplicity must be emulated in social media alternatives; using them cannot be the domain of a tech elite. Fortunately, alternative social media builders recognize this. As Alireza Mahdian et al. note, "At the current state, the trend seems to indicate that users would prefer features and convenient user experience at the cost of privacy violations. Therefore, we believe that [peer-to-peer social networks] would only be able to compete with their centralized counterparts if they can provide the same features and functionalities at the same level of user experience."[51] Similarly, TalkOpen founder XCpherX notes, "Twitter alternative is the focus of the site. And im [sic] afraid that may be taken away if we fiddle with it too much so to speak . . . just have to be careful what we do/remove/add."[52] As the authors of a technical paper on Tor note, "A hard-to-use system has fewer users—and because anonymity systems hide users among users, a system with fewer users provides less anonymity. Usability is thus not only a convenience: it is a security requirement."[53] To make any networked system viable, a critical mass of users is required, and this means coding for mass use—which in turn means drawing on existing interface conventions. To paraphrase Marx, we might build socialized media, but we do so within a specific historical context whose traditions weigh heavily on us.

Beyond the interface, much depends on activists being able to engineer a class of users for themselves; this involves guiding new users through the problems and costs of switching away from social media monopolies. Again, the heterogeneous engineers of social media recognize this: such activists as Richard Stallman (of the Free Software Foundation), Eben Moglen (of Freedombox and the Software Freedom Law Center), Jacob Appelbaum (of Tor), and Sparrow (of Crabgrass) are continuously doing the hard work of traveling conference to conference, city to city, and translating the technical details of social media alternative construction to a range of audiences. As Spideralex of Lorea notes:

> We need to engage with a relationship with our user base and the potential targeted publics. . . . We need to provide more pedagogies and didactics [so] people understand why they should use alternative social media. . . . We're not only doing it for us. . . . [W]e want to engage with people and give them alternatives and platforms that they can easily use to be more free . . . because our political ideal is that people become more free and more aware of their security and privacy.[54]

Again, this recognition is framed in relation to (although not necessarily in imitation of) dominant, proprietary software packages. The culturally ingrained us/them dichotomy that has been a part of software and technological production within capitalism—comprising We the firm and You the user—is a structure that has shaped computer culture for decades. These activists recognize that this division must be overcome, but to do so they have to operate within this dichotomy; they rely on activists from inside their organizations to bridge the gap between them and outsiders.

Of course, none of this addresses the sociopathic consumption of new electronics concomitant to increased social media use. Fortunately, there is a growing sector and history of worker-owned and nonprofit community collectives that accept donated computers, install Linux and free software on them, and provide them at a low or no cost to people who need them. Free Geek of Portland, Oregon, is a nonprofit, democratically run organization installing Linux on old computers and providing them for free to volunteers or selling them in a thrift store. Although it has ceased operation, Babbage's Basement in Ithaca, New York, was a model nonprofit that received donated computers, refurbished them, and then sold them at a low price to students and the elderly. Such projects as these demonstrate the viability of reusing machines that are deemed "obsolete" in a consumption-obsessed culture, and their ability to remain in operation for decades indicates that there is enough existing computer equipment to build a free socialized media system without resorting to buying new machines. Software developers also recognize this: developers are making small and agile systems that do not require bleeding-edge hardware; examples include the LXCE desktop systems or Puppy Linux.

Ultimately, the implementation of an idealized system, such as my proposed socialized media system, is not a simple task. In heterogeneous engineering terms, it requires an association of users, coders, volunteers at computer-recycling centers, activists, hardware, software, protocols, browsing habits, crowd-funders, messianic zeal, and press and academic coverage, among many other bits and pieces. If decentralized, peer-to-peer, antiarchival, open-source software on free hardware is the goal, then problems of organizing people and technology increase. It necessitates organization and collective effort in ways that cannot be modeled on other organizational forms, such as that of the transnational corporation. This would be unimaginable roughly two decades ago, but free and open-source production of software (such as GNU/Linux) and culture (such as in Wikipedia) can provide models. Looking over existing implementations of an idealized socialized media system, I think that despite the difficulty, there is much to be optimistic about.

Conclusion: Tools for Activists, Activist Tools

In the end, even with the production of a viable alternative to Facebook, Twitter, and Google, these tools in and of themselves do little to revolutionize anything. Rather, activists must use these tools to do their work. This is the lesson of Occupy Wall Street and the Arab Spring: social media systems are useful, but having bodies in the streets matters even more.

However, we cannot discount the embedding of particular politics within technology. The simplistic explanation of Occupy or the Arab Spring as "Facebook Revolutions" or "Twitter Movements" is based on a grain of truth: these movements would have been different—not impossible, but different—without social media. Twitter, Facebook, and Google do have positive design elements that activists have taken full advantage of, including simple interfaces, network effects, and many-to-many asynchronous communication. The fact that activists have used these systems at a very high price—that is, the price of surveillance, centralization, reduction of activism to "liking" or watching videos—reveals the need to reverse engineer current social media systems with the goal of producing better ones that can help those who do, in fact, put their bodies on the line in protests.

Thus, it is clear that the relationship between technology and activism is overdetermined. This is why it matters when people tinker and hack on technological solutions to social problems: their politics, their goals, their limitations can be replicated to a degree within the technologies they make. For example, as Sparrow of the distributed social network Crabgrass puts it, Crabgrass "really reflects as a technology object the intentions . . . and historical context of the people who originally created it. . . . We're not bashful about saying this: we're organizationally obsessed anarchists, and so we really wanted to impose on our users better organizational capacity."[55] That is to say, the makers of Crabgrass recognize that software, like any technology, contains politics within its structure, and those politics can shape use. They realize that we cannot treat software as a neutral tool; instead, we must imbue software with anticapitalist, antistate politics. Thus, the production of socialized media is the production of a common set of tools that can be ready whenever a struggle for economic and social justice erupts. As Dmytri Kleiner argues, "In order to change society we must actively expand the scope of our commons, so that our independent communities of peers can be materially sustained and can resist the encroachments of capitalism."[56] That is, we have to produce the commons and imbue them with a technologically borne politics opposed to the domination of networked capitalism.

Software can be a vital tool in this regard, because it is surprisingly obdurate: it can lie dormant, waiting in a repository or on a dusty server somewhere, waiting to be picked up, modified, and used in protest, in resistance, and as a key part of new social organizations. Obdurate socialized media software can thus be a key part of a war of positions: it certainly will not technologically determine activism—again, there is no such thing as a Twitter Revolution—but rather it can crystallize it and preserve it in software form. When they are revived and installed, if they are free and open source, well documented, and antisurveillance and have interfaces that draw on common conventions, such socialized media can be instructive in their own right, because as users work with them, they gain new perspectives on what is possible. Moreover, if politically imbued socialized media are coupled with technologically savvy activists who are also skilled in teaching others how these systems work, they can be a powerful set of tools for social justice movements. After centuries of living in capitalism, we know full well that "whatever portion of our productivity we allow to be taken from us will return in the form of our own oppression."[57] Unless, that is, we make activist tools imbued with an antioppressive politics that constantly asserts itself, tools that shape activism and that are willing to be shaped by activists. Indeed, whether those activist tools are appropriated or freely given, they help maintain the needed escape route from the noopower of social media monopolies and surveillance-based capitalism, an escape route that activists will exploit when the time is right.

Notes

INTRODUCTION

1. Available at http://www.lrntrlln.org/web2dizzaster/ (accessed October 14, 2013).
2. Schiller, *How to Think about Information.*
3. "PricewaterhouseCoopers."
4. Martinez-Torres, "Civil Society, the Internet, and the Zapatistas," 347.
5. DeLuca, *Image Politics.*
6. Froehling, "The Cyberspace 'War of Ink and Internet' in Chiapas, Mexico"; Warf and Grimes, "Counterhegemonic Discourses and the Internet"; Cleaver, "The Zapatista Effect"; Knudson, "Rebellion in Chiapas"; Bennett, "New Media Power"; Salter, "Democracy, New Social Movements and the Internet"; Kahn and Kellner, "New Media and Internet Activism."
7. Dyer-Witheford, *Cyber-Marx.*
8. Marx and Nicolaus, *Grundrisse.*
9. Agger, *Speeding Up Fast Capitalism.*
10. See Perelman, *The Perverse Economy;* Brenner, *The Boom and the Bubble.*
11. O'Reilly and Battelle, "Opening Welcome."
12. Available at http://www.facebook.com/pages/The-Chiapas-Project/300384 69340#!/pages/The-Chiapas-Project/30038469340?v=info (accessed February 2010).
13. Andrejevic, *Reality TV.*
14. Terranova, "Free Labor," 2000; Terranova, *Network Culture.*
15. Lessig, *Code.*
16. Architects, of course, have built their whole discipline on this assertion: that architecture is the "'stony' externalized memory of our human predicament,

and we constantly hit upon it" (Ebensperger, Choudhury, and Slaby, "Designing the Lifeworld," 235). This can be extended easily to the digital environment, where the very shapes, affordances, and constraints of Web software function as externalized memory, a mnemotechnics, a complex system that shapes us as much as we shape it.

17. Kitchin and Dodge, *Code/Space*, 12.

18. Berry, "The Uses of Object-Oriented Ontology"; Bogost, *Alien Phenomenology, or What It's Like to Be a Thing*.

19. Golumbia, *The Cultural Logic of Computation*.

20. Galloway and Thacker, *The Exploit;* Wark, *A Hacker Manifesto*.

21. Fuller, *Behind the Blip*.

22. Edwards, *The Closed World;* Turner, *From Counterculture to Cyberculture*.

23. Lessig, *Code*.

24. Grier, *When Computers Were Human;* Abbate, *Inventing the Internet*.

25. Kitchin and Dodge, *Code/Space*.

26. See the *Critical Code Studies* blog at http://criticalcodestudies.com/wordpress/ (accessed June 5, 2012).

27. Kirschenbaum, *Mechanisms;* see also Markley, "Boundaries."

28. Naur and Randell, "Software Engineering," 13.

29. Ceruzzi, *A History of Modern Computing*, 106.

30. *Guide to the Software Engineering Body of Knowledge*.

31. For more on this, see Gehl and Bell, "Heterogeneous Software Engineering"; Mahoney, *Histories of Computing*.

32. Along these lines, Michael Davis's article "Will Software Engineering Ever Be Engineering?" is a typical example of the denial that software production is similar to the process of engineering. Beyond this, of course, the open-source mode of software production is largely organized in a very different manner than the top-down, hierarchical software engineering model. See also Weber, *The Success of Open Source*.

33. Goffey, "Algorithm," 15–16.

34. Montfort and Bogost, *Racing the Beam,* sec. Afterword on Platform Studies.

35. Kitchin and Dodge, *Code/Space*, 13.

36. Ingle, *Reverse Engineering*, 9.

37. Chikofsky and Cross, "Reverse Engineering and Design Recovery," 15.

38. In this sense, then, reverse engineering shares many affinities with the methodology used in Kirschenbaum, *Mechanisms*.

39. Mattelart, *Networking the World, 1794–2000;* Mosco, *The Digital Sublime*.

40. Ingle, *Reverse Engineering*, 3.

41. Latour, *Reassembling the Social*.

42. Probably the best example is Mackenzie, *Cutting Code*. Adrian Mackenzie explicitly works through the associations of human and nonhuman elements/agents in his analysis of software.

43. Law, "Notes on the Theory of the Actor-Network," 380; original emphasis.

44. Law, "Technology and Heterogeneous Engineering," 111.

45. Ibid., 114.

46. Law, "Notes on the Theory of the Actor-Network," 381.

47. Of course, the very concept of an "engineer" or "organization" is itself an agent "only to the extent that matters are also decentered, unplanned, undesigned. To put it more strongly, we need to understand that to make a center is to generate and to be generated by a noncenter, a distribution of the conditions of possibility that is both present and not present" (Law, *Aircraft Stories Decentering the Object in Technoscience,* 112). In other words, the very concept of an "engineer" is fraught with tensions that could dissociate it just as a ship might be dissociated. However, despite this theoretical problem, for the purpose of this book, I am going to use such concepts as "software engineer," "new media capitalist," and "user" as useful abstractions, or tools, to perform my analysis.

48. Bogost, *Alien Phenomenology, or What It's Like to Be a Thing;* Berry, "The Uses of Object-Oriented Ontology."

49. AJAX is a Web-programming technique to provide dynamic material to client computers. On a static page, if a user makes a change (fills in a form, selects an option, etc.), the data are sent from the browser to the server. On the server side, the data are handled (often with the scripting language PHP), and then the whole package is sent back to the browser. In contrast, in AJAX, only a small part of the Web page is sent back for processing. If this is coupled with a broadband connection, the website will be more like a desktop application. For a contrast, compare a static map to a Google map. The latter can be manipulated within the frame on the page, whereas a change to the former necessitates refreshing the whole page. For more on AJAX, see Holdener, *Ajax the Definitive Guide.*

50. For a discussion of the relative lack of normative approaches to class, race, or gender in STS, see Law, *A Sociology of Monsters,* chap. introduction.

51. Marx, "Theses on Feuerbach."

52. Lovink, *Networks without a Cause,* 7.

53. Ibid.

CHAPTER 1

1. For an early description of bots (or "agents"), see Clarke, "The Digital Persona and Its Application to Data Surveillance."

2. Hwang, Pearce, and Nanis, "Socialbots," 40.

3. Boshmaf et al., "The Socialbot Network," 1, emphasis added.

4. Daw, "10 Twitter Bot Services to Simplify Your Life"; Hwang, Pearce, and Nanis, "Socialbots."

5. Boshmaf et al., "The Socialbot Network."

6. Webster, "Military's 'Persona' Software Cost Millions, Used for 'Classified Social Media Activities.'"

7. Marx, "Economic and Philosophical Manuscripts of 1844," chap. Estranged Labour.

8. Lazzarato, "The Concepts of Life and the Living in the Societies of Control."

9. See Jack Copeland's discussion in Turing, *The Essential Turing.*

10. Seife, *Decoding the Universe,* 18.

11. Turing, "Computing Machinery and Intelligence," 440.

12. The phrase "human computation" is a bit of a redundancy: "computer"

used to mean "a person who computes"; the term was later applied to electronic machines that could do the complex math that human computers could do. Today, of course, the idea of a human computer is odd and even comical. For an example, see a (very bad) 1969 Disney movie, *The Computer Wore Tennis Shoes*.

13. Grier, "Gertrude Blanch of the Mathematical Tables Project"; Grier, *When Computers Were Human;* Light, "When Computers Were Women."

14. This use of the male pronoun is wrong, and not just because it is patriarchal. At the time of Turing's writings, and indeed for many years, human computers were predominantly *women*. It was one of the few careers women with mathematical training could pursue. Women computers were involved in computation for ballistics, the Manhattan Project, and Depression-era public assistance administration.

15. Turing, "Computing Machinery and Intelligence," 438.

16. Light, "When Computers Were Women"; Grier, *When Computers Were Human.*

17. Turing, "Computing Machinery and Intelligence."

18. Turing, *The Essential Turing,* 495.

19. Turing, "Computing Machinery and Intelligence," 442.

20. Ibid.; Shannon, "The Mathematical Theory of Communication"; Wiener, *The Human Use of Human Beings.*

21. Weizenbaum, *Computer Power and Human Reason;* Turkle, *Life on the Screen.*

22. Mauldin, "Chatterbots, Tinymuds, and the Turing Test."

23. Ibid.

24. Hodges, "Alan Turing and the Turing Machine," 8; Turing, "Computing Machinery and Intelligence."

25. Davis, "Mathematical Logic and the Origin of Modern Computers," 154.

26. Hodges, "Alan Turing and the Turing Machine," 4.

27. An example of this is in Ashby, *Design for a Brain.* Ashby's emphasis is on defining discrete states for thinking and the mind.

28. And this last point is an important one. According to Webb 1980, Turing used the idea of a machine handling computable numbers by reading simple instructions to bracket off the question of an internal, unknowable intelligence within the human computer. By demonstrating that a machine could do this work just as effectively as a human computer, Turing was able to side-step the question of intelligence as inherent in an organism; after all, at the time of his early work in the 1930s, no one would consider a machine intelligent, and thus no one would accuse him of resting on a metaphysical property to explore the scope and limits of computable numbers. His only assumption regarding human intelligence is that it is finite and thus has finite "states." From here, of course, it is a short path to encoding such states in a theoretical machine, having the machine imitate those states, and then reconsidering the performance of such machines equipped as potential demonstrations of intelligence. See Webb, *Mechanism, Mentalism, and Metamathematics,* 219–225.

29. Turkle, *Life on the Screen,* 47.

30. See Golumbia, *The Cultural Logic of Computation,* for a discussion of the articulations between Chomsky's linguistics and computer science.

31. Webb, *Mechanism, Mentalism, and Metamathematics.*

32. Cybernetics is perhaps the best example, because its fundamental tenets are that biological and mechanical processes are essentially the same, that such continuous processes can be divided into discrete states, and that we can shape such processes on a system-wide (i.e., societal, evolutionary, or network) scale. Cybernetics thus enables the study of the human as machine, the regulation and improvement of human society via measuring change and monitoring feedback, and the production of equipment that can augment human capacity. Moreover, the discrete states observed in the human-machine and in society can be encoded, replicated, simulated, or calculated by Turing's Universal Machine. See Wiener, *Cybernetics;* Wiener, *The Human Use of Human Beings;* Ashby, *Design for a Brain;* Bowker, "How to Be Universal."

33. Kelly, *New Rules for the New Economy.*

34. Hodges, "Alan Turing and the Turing Machine," 5; Turing, "Computing Machinery and Intelligence," 440.

35. Deleuze, "Postscript on the Societies of Control," 4.

36. Lazzarato, "The Concepts of Life and the Living in the Societies of Control," 186.

37. Ibid., 184.

38. Ibid., 185.

39. Foucault, "The Subject and Power," 138.

40. Ibid.

41. Virtanen, "General Economy."

42. Taylor, *The Culture of Confession from Augustine to Foucault.*

43. Aldridge, "Confessional Culture, Masculinity and Emotional Work."

44. Couldry, *Media Rituals;* Andrejevic, *Reality TV.*

45. Couldry, *Media Rituals.*

46. Ibid., 125–126.

47. Andrejevic, *Reality TV,* 120–130.

48. Aldridge, "Confessional Culture, Masculinity and Emotional Work."

49. Illouz, *Cold Intimacies.*

50. Bratich, "'Nothing Is Left Alone for Too Long.'"

51. Marwick and boyd, "I Tweet Honestly, I Tweet Passionately"; Marwick and boyd, "To See and Be Seen."

52. Marwick and boyd, "To See and Be Seen," 149.

53. Gehl, "Ladders, Samurai, and Blue Collars."

54. Marwick, "Status Update."

55. See Christian, "Real Vlogs" for a discussion of "real" vlogs—that is, "sincere" or "authentic" confessional vlogs—and highly produced "fake" vlogs meant to gain audiences rather than share intimate details.

56. Marwick and boyd, "I Tweet Honestly, I Tweet Passionately," 122.

57. boyd, "Facebook's Privacy Trainwreck"; boyd, "Facebook and 'Radical Transparency' (A Rant)."

58. Terranova, *Network Culture,* chap. 5.

59. Governor, Hinchcliffe, and Nickull, *Web 2.0 Architectures,* 191, original emphasis.

60. Paolillo and Wright, "Social Network Analysis on the Semantic Web"; Liu, Maes, and Davenport, "Unraveling the Taste Fabric of Social Networks"; Liu, "Social Network Profiles as Taste Performances"; Mika, *Social Networks and the Semantic Web.*

61. Hwang, Pearce, and Nanis, "Socialbots," 40.

62. Available at http://ca.olin.edu/2008/realboy/ (accessed April 16, 2012).

63. Boshmaf, Muslukhov, Beznosov, and Ripeanu, "The Socialbot Network," 3.

64. Nanis, Pearce, and Hwang, "PacSocial," 2.

65. Boshmaf et al., "The Socialbot Network," 1.

66. See http://thetwitterbot.com/ (accessed April 8, 2012).

67. Fuglsang and Sørensen, *Deleuze and the Social,* 13.

68. Barlow, "A Declaration of the Independence of Cyberspace."

69. Lazzarato, "The Concepts of Life and the Living in the Societies of Control," 182.

70. Virtanen, "General Economy," 229.

71. Nanis, Pearce, and Hwang, "PacSocial," 1.

72. Hwang, Pearce, and Nanis, "Socialbots," 41.

73. For the text of the original solicitation, see http://wiki.echelon2.org/wiki/Persona_Management (accessed March 4, 2013).

74. Morozov, "Muzzled by the Bots."

75. Hwang, Pearce, and Nanis, "Socialbots," 44.

76. Hochschild, *The Managed Heart,* 6.

77. As Deleuze notes, "We are taught that corporations have a soul, which is the most terrifying news in the world" ("Postscript on the Societies of Control," 6).

78. Virtanen, "General Economy," 69.

CHAPTER 2

1. "The Face behind Facebook."

2. Derrida, *Archive Fever.*

3. Foucault, *The Order of Things;* Foucault, *The Archaeology of Knowledge; and, The Discourse on Language.*

4. Manoff, "Theories of the Archive from Across the Disciplines."

5. Bowker, *Memory Practices in the Sciences.*

6. Eckert, "Disclosure of a Magnetic Calculating Machine"; von Neumann, "First Draft of a Report on the EDVAC."

7. Technically speaking, Maurice Wilkes's EDSAC was the first operational stored-program computer, beginning operation two years prior to EDVAC. However, the plan for the EDVAC was the first time a practical stored-program computer was proposed, and parts of EDVAC were demonstrated to a small group of observers prior to the EDSAC. In addition, von Neumann was the first to broadly describe the stored-program concept, but credit is due to J. Presper Eckert and John Mauchly for its invention and implementation. Von Neumann gets perhaps too much credit for the concept, but for my purposes here, I follow convention and refer to the stored-program/processor architecture as the Von Neumann Architecture.

8. Grier, *When Computers Were Human;* Light, "When Computers Were Women."

9. Babbage, *On the Economy of Machinery and Manufactures, of 1835.*

10. Grier, "Gertrude Blanch of the Mathematical Tables Project," 18.

11. Von Neumann, "First Draft of a Report on the EDVAC," 11–19.

12. For a discussion of the managerial and mathematical acumen it takes to direct a room full of human computers as well as a basic discussion of the division of labor among human computers and how that division was replicated in electronic computers, see Grier, "Gertrude Blanch of the Mathematical Tables Project."

13. Aspray, *John von Neumann and the Origins of Modern Computing,* 33.

14. A History of Modern Computing.

15. Ibid., 286.

16. Backus, "Can Programming Be Liberated from the von Neumann Style?" August 1978; Backus, "Can Programming Be Liberated from the von Neumann Style?" 2007; Cantoni and Levialdi, "Matching the Task to an Image Processing Architecture (for Overcoming Von Neumann Bottleneck)"; Dickinson, *An Optical Respite from the Von Neumann Bottleneck;* Naylor and Runciman, "The Reduceron."

17. Mediati, "Windows 7 Performance Tests."

18. Mediati, "Windows 7."

19. Aspray and Campbell-Kelly, *Computer,* 92–93.

20. Indeed, despite many years of popular and academic writing on how fragile and ephemeral computer storage is, as Matthew Kirschenbaum demonstrates, contemporary computer storage is, in fact, highly durable, even surviving such disasters as the World Trade Center attacks of September 11, 2001. Here are data shifted out of time, out of one historical context and into another. See Kirschenbaum, *Mechanisms.*

21. See *Coast Artillery Field Manual.*

22. O'Reilly, "What Is Web 2.0."

23. "Traditional Ways of Judging 'Quality' in Published Content Are Now Useless."

24. Gerben, "Privileging the 'New' in New Media Literacy."

25. "The Poverty of Networks," 367.

26. "Learning to Immaterial Labour 2.0," 101.

27. "The Mortality of the Virtual."

28. Ibid., 60.

29. *The Art of the Motor,* 141; original emphasis.

30. Governor, Hinchcliffe, and Nickull, *Web 2.0 Architectures,* 127.

31. Holdener, *Ajax the Definitive Guide.*

32. "Can You Hear Me Now?"; this analysis of "information overload" and the reduction in time to contemplate is not new; see Klapp, "Meaning Lag in the Information Society."

33. Turkle, "Can You Hear Me Now?"

34. "It's Not Information Overload."

35. *Speeding Up Fast Capitalism.*

36. "Speed and Politics"; *The Art of the Motor; Negative Horizon.*

37. Terranova, "Free Labor," 48.

38. "Clickworkers on Mars."

39. *The Wealth of Networks*, 69.

40. The latest iteration of Clickworkers can be found at NASA's "Be a Martian" site, available at http://beamartian.jpl.nasa.gov/welcome (accessed December 3, 2013). "Be a Martian" allows volunteers to tag photos from Mars.

41. This project drew a lot of comparison and inspiration from a method of using the Internet to capture spare machine-computing cycles. SETI@home, a screensaver program for PCs, uses idle, networked personal computers to process the data from the Search for Extra-Terrestrial Intelligence. Started in 1999 (see http://setiathome.berkeley.edu/), this project is the "largest distributed computing effort with over 3 million users." However, in the case of Clickworkers, the massive aggregation of *human* computing/processing of data hearkens to the original meaning of the word "computer": a human who does calculations. In this case, a supercomputer is a coordinated collection of thousands or millions of such humans.

42. *The Wealth of Networks;* "Coase's Penguin, or, Linux and 'The Nature of the Firm'"; with Nissenbaum, "Commons-Based Peer Production and Virtue."

43. "The Rise of Crowdsourcing"; *Crowdsourcing.*

44. Howe, "The Rise of Crowdsourcing."

45. *The Wealth of Networks*, 3.

46. O'Reilly, "What Is Web 2.0."

47. "It's Not Information Overload."

48. This analysis is largely based on Digg as the site operated between 2004, during the early days of Web 2.0, and 2012, when the site was sold and subsequently redesigned. As of this writing, users can submit links and rate them, but they have far less control over the "new" Digg than the old. This change, which reintroduces editorial control into the user-led submission and sorting of Web content, will have to be the subject of another study.

49. Digg began as strictly a technology news aggregator meant to compete with Slashdot.org. It has since expanded to include several other categories.

50. MacManus, "Interview with Digg Founder Kevin Rose, Part 1."

51. "How Digg Works."

52. "How Digg Uncovers the News."

53. Ostrow, "Digg Bans Company That Blatantly Sells Diggs."

54. MacManus, "Interview with Digg Founder Kevin Rose, Part 1."

55. An archived version of this FAQ page is available at https://web.archive .org/web/20120208080043/http://about.digg.com/faq (accessed December 3, 2013).

56. "AI Gets a Brain."

57. "Minds for Sale."

58. Here, I have excluded Wikipedia from the "nonprofit ethic," because the original intention of Wikipedia founders Jimmy Wales and Larry Sanger was to produce a for-profit encyclopedia based on freely provided user contributions. Their hand was forced when the Spanish contingent of Wikipedia "forked" the site by setting up independent servers. This was in reaction to Wales and Sanger indicating that they would sell advertising space on Wikipedia. After this incident, Wikipedia became a nonprofit site. For further discussion, see Chapter 5.

59. Backus, "Can Programming Be Liberated from the von Neumann Style?" August 1978.

60. Cantoni and Levialdi, "Matching the Task to an Image Processing Architecture (for Overcoming Von Neumann Bottleneck)"; Backus, "Can Programming Be Liberated from the von Neumann Style?" 2007; DeBenedictis and Johnson, "Extending Unix for Scalable Computing"; Dickinson, *An Optical Respite from the Von Neumann Bottleneck;* Hartenstein, "Data-Stream-Based Computing"; Hartenstein, "The Digital Divide of Computing"; Naylor and Runciman, "The Reduceron."

61. Lyman and Varian, "How Much Information?"

62. Wray, "Internet Data Heads for 500bn Gigabytes."

63. For an explanation of this typical AJAX request, see the W3's tutorial at http://www.w3schools.com/XML/xml_http.asp.

64. Flanagin, Flanagin, and Flanagin, "Technical Code and the Social Construction of the Internet," 186.

65. Drawing on Marx, Nick Dyer-Witheford (*Cyber-Marx,* 202) explores this contradiction. While informational flows promise to speed up the Money-Commodity-Money' circuit, "for the commodity to retain its essential attribute—that of being bought and sold—its passage must be interrupted: 'it must spend some time as a cocoon before it can take off as a butterfly.' Today, electronic technologies are making a whole range of commodities central to the information economy—computer software, films, video, television programs, electronic music and games . . . into instant butterflies." Social media avoids this by using log-ins, IP, and Terms of Service to cocoon data long enough for an exchange to be possible.

66. This dialectic of timelessness and concreteness (i.e., existing in real time) is explored in Sohn-Rethel, *Intellectual and Manual Labour,* 61–64.

67. Huws, *The Making of a Cybertariat.*

68. Fisher, *Media and New Capitalism in the Digital Age,* 137–143.

69. Zuckerberg, "A Conversation with Mark Zuckerberg" (roughly the seven-minute mark).

70. Bruns, "The Future Is User-Led."

71. Anderson, *The Long Tail.*

72. For a discussion of this process in terms of YouTube, see Gehl, "YouTube as Archive"

73. Available at http://www.facebook.com/terms.php (accessed 3 December 2013); my emphasis.

74. See Bianco, "Social Networking and Cloud Computing," for a similar analysis of the Terms of Service of Flickr and YouTube.

75. Chang, "To the Archive," 204.

76. Ibid., 205.

77. "Reading an Archive," 184.

78. Bowker, *Memory Practices in the Sciences,* 18.

79. Deleuze, "Postscript on the Societies of Control."

80. This is not to say that the metrics of security and of commerce are mutually exclusive. The all-too-common state practice of knocking on the doors of ISPs and social networks and demanding records of user activity points to the usefulness of consumption data for law enforcement.

81. Lanier, *You Are Not a Gadget.*
82. Mosco, *The Digital Sublime.*
83. Fuller and Goffey, "Digital Infrastructures and the Machinery of Topological Abstraction," 327.
84. Reality TV, 18.
85. Zittrain, *The Future of the Internet and How to Stop It.*
86. For analysis of the post hoc production of evidence, see the work of Kelly Gates, particularly her book *Our Biometric Future.*
87. Gehl, "YouTube as Archive"
88. Available at https://addons.mozilla.org/en-US/firefox/addon/13993 (accessed January 1, 2010).
89. Kirschenbaum, *Mechanisms,* chap. 2.

CHAPTER 3

1. Fixmer and Rabil, "MySpace Sale, Merger or Spinoff Being Weighed by News Corp., Official Says."
2. Goldman, "MySpace to Cut 30%, or 430 Jobs."
3. Perez, "MySpace Slashes Nearly Half of Staff."
4. Web-traffic site Alexa.com ranked Myspace in the top 10 worldwide as late as 2009, but Myspace has fallen to 798 as of December 2013.TechTree India found that Facebook's traffic surpassed that of Myspace in August 2008 ("Facebook"). As another indicator, Google Trends search data as of this writing show that searches for "myspace" peaked in 2007 and have steadily declined to pre-2006 levels. Finally, according to *Daily Finance*'s Jeff Bercovici, Myspace's internal estimates of traffic and users have consistently been lower than the numbers the company has published in its press kit. The numbers in the press kit come from ComScore, a Web-metrics competitor to Alexa. This discrepancy casts doubts on precisely how much traffic Myspace has received, even in its heyday (Bercovici, "MySpace Users").
5. Boorstin, "MySpace Finally Sold for Some $35 Million."
6. Cashmore, "MySpace, America's Number One."
7. Foresman, "Myspace Concedes to Facebook, Changes Focus."
8. Raphael, "Myspace's Facebook 'Mashup'—Why Bother?"; Sullivan, "Myspace to Use Facebookers' Entertainment Likes."
9. Bijker, Hughes, and Pinch, *The Social Construction of Technological Systems.*
10. Brooks, *The Mythical Man-Month.*
11. Ibid., 45, original emphasis.
12. Ibid., 42.
13. Colburn and Shute, "Abstraction in Computer Science," 173.
14. Ibid., 177.
15. Parnas and Siewiorek, "Use of the Concept of Transparency in the Design of Hierarchically Structured Systems."
16. *Guide to the Software Engineering Body of Knowledge,* chap. 3 (available at http://www.computer.org/portal/web/swebok/html/ch3).
17. Liskov and Guttag, *Program Development in Java,* 2.
18. See also Nandhakumar, "Managing Time in a Software Factory."

19. See Phillipe Krutchen's preface in Bosch, *Design and Use of Software Architectures*, xii.

20. Brooks, *The Mythical Man-Month*, 256.

21. Sohn-Rethel, *Intellectual and Manual Labour*.

22. Marx, *A Contribution to the Critique of Political Economy*.

23. Sohn-Rethel, *Intellectual and Manual Labour*, 19.

24. Ibid., 20.

25. Toscano, "The Open Secret of Real Abstraction," 275.

26. Ibid.

27. Fuller and Goffey, "Digital Infrastructures and the Machinery of Topological Abstraction," 321.

28. Marx, *Grundrisse*, 164.

29. Of course, software need not be developed in such a rigidly hierarchical system. Much nonproprietary open-source software has been produced in more egalitarian and playful ways.

30. Raymond, *The Cathedral and the Bazaar*.

31. O'Reilly, "What Is Web 2.0"; O'Reilly et al., "The Architecture of Participation."

32. Benkler, *The Wealth of Networks*; Bruns, *Blogs, Wikipedia, Second Life, and Beyond*.

33. O'Reilly, "What Is Web 2.0"; O'Reilly et al., "The Architecture of Participation."

34. That said, it would be a mistake to let Web-based software drift off into a hazy conceptual cloud: such software *must* materialize, and it does so in great part on physical, very real server farms. These systems are painfully material, drawing massive amounts of power and thus contributing to energy crises and carbon emissions. The common mistake here, of course, is to simply assume this materiality away on the basis of individual experience with, say, a smartphone or a Google Chromebook.

35. Howe, *Crowdsourcing*.

36. Governor, Hinchcliffe, and Nickull, *Web 2.0 Architectures*, xi.

37. Andrejevic, *Reality TV*; Andrejevic, *iSpy* Andrejevic's analysis of "interaction" reveals its reductive form, where "participation" is often limited to making consumer choices.

38. Scharman, "'You Must Be Logged in to Do That!'"; Scholz, "Market Ideology and the Myths of Web 2.0."

39. Coté and Pybus, "Learning to Immaterial Labour 2.0," 96.

40. Available at http://www.facebook.com/pages/Satanist/110943382261091 (accessed June 1, 2012).

41. Available at http://www.myspace.com/pissandshitandsugar. My description is based on a viewing of this page before Myspace was sold. Sometime after the site was sold, this page for Satan was altered, and the current version appears to be far more sedate. For screenshots of presale, pimped Myspace pages, feel free to e-mail me.

42. This image is still online, available at https://a3-images.myspacecdn.com/images03/33/98056838f5784347ad5b4e212bb6a17b/full.jpg (accessed December 4, 2013).

43. boyd, "Viewing American Class Divisions through Facebook and My-Space"; boyd, "Facebook and MySpace Users Are Clearly Divided along Class Lines."

44. Countless websites offer code to remove Myspace ads, although most of them note that doing so is a violation of the Myspace Terms of Service. For examples, see http://www.myspacelayoutsupport.com/myspace-tutorials/removing-myspace-ads.php; http://answers.yahoo.com/question/index?qid=2007060715522 7AAUFlCT; http://abrax.us/Katamari/RemoveLayoutAds.php.

45. To be certain, there are many ways to block ads in Facebook, including such browser extensions as Adblock Plus. However, these extensions block ads from appearing only to individual users who use these extensions; the ads are still there for others. The code specific to Myspace, which would made ads invisible to *anyone* who visited a pimped Myspace profile, pointed to user desire to control advertising at the point of production rather than consumption, which is a substantially different form of resistance to marketing. Rather than block advertisements while they use the Web, those who attempt to block advertisements on their own profiles are attempting to prevent their profiles from being commodified.

46. Vega and Kopytoff, "Web Advertisers Fear Effects of Do-Not-Track System."

47. boyd, "Friendster and Publicly Articulated Social Networking"; boyd and Ellison, "Social Network Sites."

48. Poulsen, "MySpace Predator Caught by Code."

49. de Vries, "MySpace Tightens Age Restrictions."

50. Mattus, "Legal Update"; Kay, "Extending Tort Liability to Creators of Fake Profiles on Social Networking Websites."

51. Szoka and Thierer, "Cyberbullying Legislation"; Turbert, "Faceless Bullies"; Whitaker and Bushman, "Online Dangers"; Willard, "Cyberbullying Legislation and School Policies."

52. boyd and Jenkins, "MySpace and Deleting Online Predators Act (DOPA)"; Marwick, "To Catch a Predator? The MySpace Moral Panic."

53. There appear to be no academic studies of fake profiles on Facebook yet. However, many users have filed reports on the Facebook Governance page, claiming to have had their identities stolen: see http://www.facebook.com/topic.php?uid=69178204322&topic=19372. In addition, it is easy to set up a fake profile on Facebook, but it is difficult to make friends with such a profile because of the culture of the site.

54. Yadav, "Facebook—The Complete Biography"; Phillips, "A Brief History of Facebook"; *Social Network Websites.*

55. Cuban, "An Open Facebook API Vs Google OpenSocial."

56. Optimize IT, "Facebook Wants to Issue Your Internet Driver's License"; Howard, "2011 Trends."

57. O'Malley, "MySpace Blossoms into Major Web Portal."

58. Available at http://www.myspace.com/imolovinoit (accessed March 18, 2011).

59. Morrissey, "MySpace Is Still Here."

60. Constine, "Myspace Lost $43M in 2012."

61. Shields, "Facebook Trumps MySpace on Ads."

62. BBC News, "Facebook Revenue Surges 60% on Strong Ad Sales," October 31, 2013 (available at http://www.bbc.co.uk/news/business-24751441).

63. Spolsky, "The Law of Leaky Abstractions."

64. Rosenberg, *Dreaming in Code.*

65. Tynan, "The 25 Worst Web Sites."

66. Illouz, *Cold Intimacies,* 4.

67. Sohn-Rethel, *Intellectual and Manual Labour,* 30.

68. Ibid.

69. Ibid., 33.

70. Whitehead, *Science and the Modern World,* 59.

71. Leistert, "How to Put Up a Facebook Resistance."

72. Zuckerberg's mission to connect the world to Facebook and to make everything open and transparent can be seen in Facebook's SEC S-1 filing statement (available at http://www.sec.gov/Archives/edgar/data/1326801/000119312512034517/d287954ds1.htm) as well as his Internet.org project. In his public presentations of his projects, he consistently declares that the only way to rid the world of problems is to connect everyone to another—no secrets, no anonymity, no hiding. Even if one is not concerned about this language, consider the recent revelations that Facebook has patented a means to efficiently transfer data to government agencies, such as the NSA (see http://patft.uspto.gov/netacgi/nph-Parser?Sect1=PTO1&Sect2=HITOFF&d=PALL&p=1&u=%2Fnetahtml%2FPTO%2Fsrchnum.htm&r=1&f=G&l=50&s1=8,438,181.PN.&OS=PN/8,438,181&RS=PN/8,438,181). Facebook's success in building a centralized social networking site makes it a prime target for state agencies.

73. boyd, "Facebook's Privacy Trainwreck."

74. Gehl, "'Why I Left Facebook.'"

CHAPTER 4

1. This is in spite of the fact that they have been dubbed a key object of study for the Society of People Interested in Boring Things. See Star, "The Ethnography of Infrastructure"; Lampland and Star, *Standards and Their Stories.*

2. Townes, "The Spread of TCP/IP and the Political Origins of the Internet."

3. O'Reilly, "The Architecture of Participation."

4. Rimmer, "Napster"; Benkler, *The Wealth of Networks;* Bruns, *Blogs, Wikipedia, Second Life, and Beyond.*

5. Raymond, *The Cathedral and the Bazaar;* Lee and Cole, "From a Firm-Based to a Community-Based Model of Knowledge Creation."

6. Berman, "Iran's Twitter Revolution"; "Iran's Twitter Revolution."

7. Arola, "The Design of Web 2.0."

8. Russell, "'Industrial Legislatures,'" 1.

9. Egyedi, "Consortium Problem Redefined," 28.

10. Egyedi, "Consortium Problem Redefined," 26.

11. Swann, "The Economics of Standardization," vi–vii.

12. See Mattelart, *Networking the World, 1794–2000;* Noble, *America by Design;* Russell, "'Industrial Legislatures.'"

13. Hawkins, "The Rise of Consortia in the Information and Communication Technology Industries," 160.

14. As Hawkins (ibid., 162) notes, the European Computer Manufacturer Association (ECMA), which was the first standards consortium (founded in 1963) set the model for many consortia to follow, including voluntary, industry-wide international membership; the use of technical committees; funding via membership dues; and the open publication of its standards.

15. Ibid., 161.

16. Of course, the process of producing Internet standards is not divorced from state aims. The Internet Corporation for Assigned Names and Numbers (ICANN) has been consistently criticized for being under the thumb of the U.S. government. Setting this issue aside, the dominant ideology of Internet standards is of a freewheeling, libertarian process that is technologically neutral—that is, standards arise simply from one criterion: does the technology work? For a more nuanced and critical analysis of this process, see Janet Abbate's history of the Internet, *Inventing the Internet*, and Alexander Galloway's work on protocol, *Protocol*.

17. Swann, "The Economics of Standardization," iii, v.

18. Weiss and Cargill, "Consortia in the Standards Development Process," 560.

19. Egyedi, "Consortium Problem Redefined," 33.

20. Weiss and Cargill, "Consortia in the Standards Development Process," 561.

21. Ibid.; Christ and Slowak, "Why Blu-Ray vs. HD-DVD Is Not VHS vs. Betamax."

22. Boh, Soh, and Yeo, "Standards Development and Diffusion," 58.

23. Weiss and Cargill, "Consortia in the Standards Development Process"; Swann, "The Economics of Standardization."

24. In Swann's "The Economics of Standardization," an economic analysis of standards consortia, he finds that consumer groups are often left out of the standards-setting process, even though standards can affect them significantly.

25. Hawkins, "The Rise of Consortia in the Information and Communication Technology Industries," 170.

26. In fact, as Hawkins (ibid., 164) notes, private standards consortia would have largely run afoul of U.S. antitrust legislation until the law was changed in 1988: "Changes in US laws from 1988 onwards encouraged closer R&D collaboration between US firms, and it is against this background that most of the current generation of consortia emerged." This shift in antitrust legislation is, of course, part of the larger move toward deregulation of corporations in the West.

27. See McChesney, *The Political Economy of Media*.

28. Dreze and Zufryden, "Is Internet Advertising Ready for Prime Time?"

29. For academic discussions of this phenomenon, see Benkler, *The Wealth of Networks*; Bruns, *Blogs, Wikipedia, Second Life, and Beyond*. For a discussion of "prosumption" from a marketing perspective, see Vollmer, *Digital Darwinism*.

30. Wells and Chen, "The Dimensions of Commercial Cyberspace." It is unwise to overstate the "chaos" of the early Web; it was no more "chaotic" than, for example, early radio. The production of the idea of "chaos" in a new medium is

often a precursor to intense regulation, both by states and by corporations. This has happened with radio and is currently happening with the Web. For discussions of this process in radio, see Douglas, *Inventing American Broadcasting, 1899–1922;* McChesney, "Conflict, Not Consensus."

31. For examples, see Jack Lynch's speech about the Web (available at http://andromeda.rutgers.edu/~jlynch/Papers/web.html), Cynthia Waddell's discussion of accessibility (available at http://corporate.findlaw.com/law-library/applying-the-ada-to-the-internet-a-web-accessibility-standard.html), and Richard Balough's discussion of e-commerce in the 1990s (available at http://www.balough.com/WorkArea/showcontent.aspx?id=1246).

32. Collins, "Trade Groups Propose Web Banner Guidelines."

33. Zuckerman, "Firms Competing to Define Web Demographics"; Cleland, "Media Buying and Planning."

34. Williamson, "Internet Ad Bureau to Tally Ad Spending."

35. "IAB Misstep"; see also Stross, *The Microsoft Way,* for a discussion of how Microsoft viewed the advent of the Internet largely as a possible "interactive television" system.

36. At that time, the IAB was called the Internet Advertising Council. Shortly after, it changed to the Internet Advertising Bureau, then later to the Interactive Advertising Bureau. For simplicity's sake, I refer to it by its current name.

37. Interactive Advertising Bureau, "Voluntary Guidelines for Banner Advertising and Process for Exploring Future Internet Advertising Forms Announced by Industry Groups"; Williamson, "Groups Set to Unveil Web Ad Guidelines."

38. IAB Pop-Up Task Force, "Pop-Up Guidelines and Best Practices."

39. IAB, "Ad Unit Guidelines."

40. IAB, "Rich Media Creative Guidelines."

41. IAB, "IAB Announces Winners of 'Rising Stars' Competition; New Brand-Friendly Ad Formats Aim to Spur Greater Creativity in Interactive Advertising"; Creegan, "Dodge Becomes First Advertiser to Tap Microsoft's IAB Award-Winning Filmstrip."

42. For a discussion of the difficulty of integrating pre-IAB standard advertising into website design, see DiNucci, *Elements of Web Design,* 65.

43. See Bernard, "Developing Schemas for the Location of Common Web Objects."

44. Granka, Hembrooke, and Gay, "Location Location Location."

45. Benway, "Banner Blindness"; Pagendarm and Schaumburg, "Why Are Users Banner-Blind?"; Benway and Lane, "Banner Blindness"; Burke et al., "Banner Ads Hinder Visual Search and Are Forgotten"; Chatterjee, Hoffman, and Novak, "Modeling the Clickstream"; Bernard, "Developing Schemas for the Location of Common Web Objects."

46. Bruner, *The Decade in Online Advertising,* 5.

47. Foucault, *Discipline and Punish,* 153.

48. In 1996, Web advertising was heavily concentrated in portal sites, such as Yahoo!, Lycos, Excite, and Infoseek (Akin, "Casting the Net"). The portal model was a mass media style of Web publishing and advertising, where publishers such as Yahoo! would create a wide range of content for as large an audience as possible.

A major reason why marketing was able to break out of the portal model in favor of the more distributed Web 2.0 model was the advent of metrics and forms of tracking that could be used across multiple sites.

49. Williamson, "Interactive"; IAB, "IAB Publishes Glossary of Terms"; IAB, *Glossary of Interactive Advertising Terms V. 2.0;* See also the IAB–United Kingdom document, *IAB Jargon Buster.*

50. IAB, "IAB Publishes Glossary of Terms."

51. IAB and American Association of Advertising Agencies, *Standard Terms and Conditions for Internet Advertising for Media Buys One Year or Less (Version 3.0).*

52. Zuckerman, "Firms Competing to Define Web Demographics."

53. PricewaterhouseCoopers, *IAB Online Ad Measurement Study.*

54. Ibid., 5.

55. IAB, "IAB Issues First-Ever Guidelines for Interactive Audience Measurement and Advertising Campaign Reporting and Audits."

56. The full list of "General Members," which are "corporate entities or stand-alone divisions of a corporate entity whose revenue is significantly based on the sale of interactive advertising inventory," is available at http://www.iab.net/member_center/1521 (accessed May 23, 2012).

57. A relatively recent example was Microsoft's protest to the Justice Department over Google's purchase of the ad network DoubleClick. Microsoft, Yahoo!, and Time Warner failed in their bids to buy DoubleClick. All these companies are General Members of the IAB. Thus, despite the façade of Schumpeterian struggle presented in the tech reporting, the situation might be more akin to Adam Smith's famous observation: "People of the same trade seldom meet together, even for merriment and diversion, but the conversation ends in a conspiracy against the public, or in some contrivance to raise prices. It is impossible indeed to prevent such meetings, by any law which either could be executed, or would be consistent with liberty or justice" (*Wealth of Nations,* book 1, chap. 10, para. 82).

58. O'Reilly, "What Is Web 2.0," 19.

59. Desilva and Phillips, LLC, *Online Ad Networks.*

60. For a discussion of Facebook's and Google's constant design experimentation conducted to optimize the amount of user data inputted, see Andrejevic, "Surveillance and Alienation in the Online Economy."

61. Vollmer, *Digital Darwinism,* 12.

62. Andrejevic, *iSpy;* Sparrow, "Social Media and the Rise of Surveillance-Based Advertising."

63. Vollmer, *Digital Darwinism,* 7.

64. The way beacons work involves third-party ad servers linked to publisher sites. For example, say the *Washington Post, New York Times,* and *Miami Herald* all contract with the same third-party ad network. That ad network inserts the same 1×1-pixel invisible GIF image into each of those papers' sites. A user who visits each site will trigger the same image from the same server every time, thus allowing the ad server to know that the visitor went to those sites. With the addition of tracking software, the ad server also learns which articles the user read, which ads he or she saw, and even which search terms the user typed in. See Center for Digital Democracy, U.S. Public Interest Research Group, and World Privacy

Forum, *Complaint, Request for Investigation, Injunction, and Other Relief,* 6; see also Richard Smith's discussion in *Ehavorial Advertising,* 14–34.

65. Galison, "The Ontology of the Enemy."

66. Benninger, "AJAX Storage," 2 3.

67. For a more recent discussion of these metrics as they apply to social media, see IAB User-Generated Content and Social Media Committee, *Social Media Ad Metrics Definitions.*

68. For a startling demonstration of this, see the Electronic Frontier Foundation's "Panopticlick" project (available at http://panopticlick.eff.org/). See also Eckersley, "How Unique Is Your Web Browser?"

69. boyd, "Friendster and Publicly Articulated Social Networking," 1.

70. Although the "Semantic Web" is a complex creation, built on standardized metadata associated with digital objects, as Peter Mika (Mika, *Social Networks and the Semantic Web,* 1:24) notes, in Web 2.0 users are capable of producing metadata (often in the form of folksonomies) as long as this work is presented in a user-friendly and abstracted manner.

71. In fact, it appears that both Facebook and Google+ are hoping that such intermingling of user and brand activities becomes so enculturated in users that the new advertisements featuring profile pictures of users as endorsers will be accepted. For example, see Sengupta, "So Much for Sharing His 'Like'"; Learmonth, "Google to Add +1 Social Layer to Display Ads."

72. Mika, *Social Networks and the Semantic Web,* 24; original emphasis.

73. Interestingly, "Scott Hughes" is himself a marketing fiction, a persona created by Axciom to promote its services. In this sense, Hughes represents an ideal standard by which to measure human activity. Axciom's customers can measure the quality of its service by the extent to which humans replicate Hughes's putatively serendipitous path from seeing a Facebook friend "like" a product to looking it up on the retailer's site to seeing an ad for it the next day (and perhaps not even connecting the appearance of the ad to his previous activity) to ultimately buying it. Hughes thus performs much like the socialbots described in Chapter 1. Hughes might well be considered a specification for a socialbot, and thus, he becomes a specification for the ideal actions of consumers online.

74. Singer, "Acxiom, the Quiet Giant of Consumer Database Marketing."

75. IAB User-Generated Content and Social Media Committee, *Social Advertising Best Practices,* 7.

76. As Elijah Sparrow ("Social Media and the Rise of Surveillance-Based Advertising," 5) astutely notes, "walled gardens" are antithetical to the overall ideology of Web 2.0: "The first impulse of many Internet companies is to guard the information on which their business depends. According to the web 2.0 model, this impulse is exactly wrong.Instead, data which is locked up in a 'walled garden' or a 'data silo' should be set free. By embracing open, standards-based protocols and allowing their data to be used by other people and companies, a web 2.0 company is able to harness the entire Internet and connect its service with everyone else, providing mutually beneficial value." In the case of social networking sites, the standardized protocols center on the production of user datasets to be "used" by marketing firms. See also a discussion of "walled gardens"

as an "anti–Web 2.0 pattern" in Governor, Hinchcliffe, and Nickull, *Web 2.0 Architectures*, 182.

77. boyd, "Viewing American Class Divisions through Facebook and My-Space"; boyd, "Facebook and MySpace Users Are Clearly Divided along Class Lines."

78. For a detailed explanation of the Facebook API, see Hawker, *The Developer's Guide to Social Programming*.

79. To be sure, given the spectacular amounts of capital flowing through social media, one wonders whether we are heading for a second Internet bubble. Google's purchase of YouTube for $1.65 billion in 2006 has not paid off; YouTube still does not turn a profit. The massive valuation of Facebook and LinkedIn (both at around $50 billion—more than many industrial and manufacturing firms) strikes me as unrealistic as tiny condos selling for a half-million dollars in 2007. Facebook's IPO was plagued by software glitches and fears that the social network has hit a wall. Zynga's stock price is falling. Other tech IPOs have been put on hold.

80. Bennett, *The Privacy Advocates*.

81. Smith, *Discerning the Subject*, xxix.

82. Ibid., 5.

83. For a discussion of how social sciences (such as modern marketing) privilege their abstract objects of study while displacing heterogeneity, see Law, "Making a Mess with Method."

84. Beck, *Risk Society;* Giddens, *Modernity and Self-Identity;* Adams, "Sourcing the Crowd for Health Services Improvement."

85. Bruns, *Blogs, Wikipedia, Second Life, and Beyond.*

86. Gehl, "Ladders, Samurai, and Blue Collars."

87. Andrejevic, *iSpy*, 29.

88. The other members of this alliance, which became the Digital Advertising Alliance (DAA) in 2010, are the 4As, the Network Advertising Initiative, the Association of National Advertisers, the Council of Better Business Bureaus, and the Direct Marketing Association. I have focused on the IAB here because of its explicit mission to ward off adverse legislation, a mission that is more implicit in the other organizations. Properly speaking, however, the AboutAds.info site, detailed below, is the creation of the DAA as a whole.

89. IAB et al., *Self-Regulatory Principles for Online Behavioral Advertising.*

90. Available at http://www.iab.net/privacymatters/ (accessed May 22, 2012).

91. An example of this logo—enlarged to show texture—is available at http://www.aboutads.info/ (accessed June 1, 2012).

92. Available at http://www.aboutads.info/choices (accessed May 22, 2012).

93. This is not scientific, but every time I have visited this site—and I have done so often throughout my research on the IAB—the opt-out process failed for a percentage of the companies involved, ranging between 10 percent to 50 percent. My findings are corroborated by a detailed study by Komanduri et al. ("AdChoices?" 629).

94. Specifically, at http://www.aboutads.info/enforcement/ (accessed May 22, 2012).

95. Maxwell and Miller, *Greening the Media,* 14.

96. For an analysis of how the similar discourses of self-regulation work in other sectors of the economy, see Nadesan, "Transparency and Neoliberal Logics of Corporate Economic and Social Responsibility."

97. As it has in the past. For example, the transition of the DNS from the U.S. government to the private ICANN was marked by discourses of "self-regulation." Similar to what I am exploring here, Milton Mueller argues that such discourses cover up the politics and economics of such a transition; see Mueller, "ICANN and Internet Governance."

98. *Consumer Data Privacy in a Networked World;* Federal Trade Commission, *Self-Regulatory Principles for Online Behavioral Advertising.*

99. Galloway and Thacker, *The Exploit,* 29.

100. Gehl, "Distributed Centralization."

101. Galloway, *Protocol,* 95.

102. See Galloway and Thacker, *The Exploit,* 81.

103. Bratich, "User-Generated Discontent," 624.

104. Shapiro, "Obama, Romney Campaigns Taking 'See What Sticks' Approach to Web Videos."

105. Cohen, "An Ad Blocker Opens the Gate, Ever So Slightly."

106. Ghostery is owned by Evidon, a company that helps online advertisers meet their self-regulatory guidelines. Ghostery allows users to participate in "GhostRank," which tracks which ads are blocked, where they appear, and when they were encountered by users.

107. Galloway and Thacker, *The Exploit,* 84.

108. See the *Wall Street Journal*'s series "What They Know," available at http://online.wsj.com/search/aggregate.html?article-doc-type={What+They+Know}.

CHAPTER 5

1. Terranova, "Free Labor," 2000; Terranova, *Network Culture.*

2. Postigo, "Emerging Sources of Labor on the Internet"; Postigo, "Of Mods and Modders"; Postigo, "America Online Volunteers."

3. Zwick, Bonsu, and Darmody, "Putting Consumers to Work."

4. Andrejevic, *Reality TV;* Andrejevic, *iSpy.*

5. Terranova, "Free Labor," 33.

6. Marwick, "Status Update," 13.

7. For a discussion of the market for Twitter followers as well as the reasons why someone might buy a following, see Considine, "Twitter Followers for Sale."

8. For an early analysis of the tension between amateur and professional production in mass media, see Enzensberger, "Constituents of a Theory of the Media," 76–77.

9. Robins and Webster, "The Revolution of the Fixed Wheel."

10. Hardt and Negri, *Empire;* Hardt and Negri, *Multitude.*

11. Dean, "The Networked Empire."

12. Woolgar, "Configuring the User."

13. Gillespie, "Designed to 'Effectively Frustrate.'"

14. Fisher, *Media and New Capitalism in the Digital Age,* 137–143.

15. Finelli, *Astrazione e Dialettica Dal Romanticismo Al Capitalismo;* Rancier, "The Concept of 'Critique' in the 'Critique of Political Economy'"; Toscano, "The Open Secret of Real Abstraction."

16. Quoted in Toscano, "The Open Secret of Real Abstraction," 273.

17. Ibid., 279.

18. Sohn-Rethel, *Intellectual and Manual Labour.*

19. Dugdale, "Materiality," 118.

20. Marx, *The Poverty of Philosophy.*

21. Both the Twitter and Facebook SEC Filing Statements for their IPOs are replete with this language: monthly active users, daily active users, total Internet users. Public investment in the social network hinges on its scale and thus the number of people who can be reached via advertisements. In the case of the Facebook S-1 statement, there is far less discussion of what can actually be done on Facebook, beyond abstract ideas, such as "democratic communication." See Ebersman, "Facebook, Inc. Registration Statement on Form S-1"; Costolo, "Twitter, Inc., Registration Statement on Form S-1."

22. Marx, *The Poverty of Philosophy.*

23. Bottero, "Class Identities and the Identity of Class."

24. Burawoy, "Reflections on the Class Consciousness of Hungarian Steelworkers," 25.

25. Hall, "Gramsci's Relevance for the Study of Race and Ethnicity," 14.

26. These criteria are drawn from Wright, "Foundations of a Neo-Marxist Class Analysis," 23.

27. For a discussion of this form of dialectical antagonism, see the footnote in ibid., 25.

28. Lih, *The Wikipedia Revolution,* 21.

29. Ibid., 31.

30. Postigo, "America Online Volunteers."

31. Lih, *The Wikipedia Revolution,* 40–41.

32. Szattler, "GPL Viral License or Viral Contract."

33. According to Lih (*The Wikipedia Revolution,* 72), the Nupedia used a "Nupedia Open Content License, which allowed people to copy and modify content. However, with that license Bomis, Inc., was still the legal copyright holder."

34. Famiglietti, "Hackers, Cyborgs, and Wikipedians," 111.

35. Lih, *The Wikipedia Revolution,* 63–64.

36. Jochym, "Re: [Intlwiki-l] Test the New Software!"

37. "My Views on the International Wikipedias and What They Should Do."

38. Enyedy, "Spanish Wikipedia."

39. Famiglietti, "Hackers, Cyborgs, and Wikipedians," 161; Tkacz, "The Politics of Forking Paths," 102.

40. "30 Jan Stats."

41. Of course, Spanish is the de facto second language of the United States. However, the Spanish version of Wikipedia was led by citizens of Spain.

42. Enyedy, "Five Messages," February 19, 2002.

43. Sanger, "Announcement about My Involvement in Wikipedia and Nupedia."

44. Enyedy, "Good Luck with Your wikiPAIDia."

45. Enyedy and Tkacz, "'Good Luck with Your WikiPAIDia.'"

46. Ibid.

47. In particular, see this thread: Wegrzanowski, "Polish Wikipedia and PHP." In 2002, Wegrzanowski repeatedly asked for a test site for the Polish Wikipedia, but answers to these requests appeared to be slow in coming.

48. Many of these documents are in Spanish; English translations are my own.

49. Enyedy, "Five Messages," February 19, 2002.

50. Sanger, "Re: [Intlwiki-l]."

51. Enyedy and Tkacz, "'Good Luck with Your WikiPAIDia.'"

52. "Por Qué Estamos Aquí y No en Es.wikipedia.org."

53. Ruiz Rivas, "Usuario."

54. "Enciclopedia Discusión."

55. Wright, "A General Framework for the Analysis of Class Structure," 16–17.

56. Weber, *The Success of Open Source;* Famiglietti, "Hackers, Cyborgs, and Wikipedians"; Tkacz, "The Politics of Forking Paths."

57. See Alperovitz, *America beyond Capitalism*, for a discussion of this distinction.

58. Sanger, "Why Wikipedia Must Jettison Its Anti-elitism."

59. Famiglietti, "Hackers, Cyborgs, and Wikipedians," 160.

60. Tkacz, "The Politics of Forking Paths," 104.

61. Cohen, "In Google-Verizon Deal, Fears for Privacy."

62. Andrejevic, "Surveillance and Alienation in the Online Economy," 280.

63. Fuchs, "The Political Economy of Privacy on Facebook," 148.

64. Kleiner, *The Telekommunist Manifesto,* 10.

65. Albrechtslund, "Online Social Networking as Participatory Surveillance."

66. Available at http://wikiscanner.virgil.gr/ (accessed May 4, 2012).

67. A comparison is available at http://en.wikipedia.org/w/index.php?diff=prev&oldid=8931861 (accessed May 4, 2012). The IP address is 192.67.48.156 and owned by Exxon-Mobil.

68. This comparison is available at http://en.wikipedia.org/w/index.php?title=Exxon_Valdez_oil_spill&diff=next&oldid=8931861 (accessed 4 May 2012).

69. Available at https://en.wikipedia.org/w/index.php?title=Wikipedia:No_original_research&oldid=487729228 (accessed April 23, 2012).

70. Available at https://en.wikipedia.org/w/index.php?title=Wikipedia:Verifiability&oldid=484523864 (accessed March 3, 2012).

71. Available at https://en.wikipedia.org/w/index.php?title=Wikipedia:Neutral_point_of_view&oldid=487851552#Due_and_undue_weight (accessed May 10, 2012).

72. To read more, see Messer-Kruse, "The 'Undue Weight' of Truth on Wikipedia"; Rosen, "Does Wikipedia Have an Accuracy Problem?"

73. Enyedy and Tkacz, "'Good Luck with Your WikiPAIDia.'"

74. Glott, Schmidt, and Ghosh, "Wikipedia Survey—Overview of Results."

75. Famiglietti, "Hackers, Cyborgs, and Wikipedians."

76. Tkacz, "The Politics of Forking Paths," 106.

77. Dugdale, "Materiality."

CHAPTER 6

1. Ingle, *Reverse Engineering*, 2.
2. Moody, *I Sing the Body Electronic*.
3. Enzensberger, "Constituents of a Theory of the Media."
4. Berners-Lee and Fischetti, *Weaving the Web*.
5. For a discussion of this troubling gap between upload and download speeds, see Galloway, *Protocol*, 68, note 25.
6. Deep Packet Inspection (DPI) is a technology that allows a network operator to discover the contents of individual packets of data flowing through a network node. So, rather than simply looking at the header contents of the packet, which contain addressing and reassembly information, a network operator looks to see whether the packet of data is, say, part of an e-mail, part of a Skype call, part of a video file, or part of a BitTorrent file. There are many disconcerting aspects of this, two of which are very important. First, ISPs can use DPI to slow down content they do not approve of, thus changing the largely neutral Internet into a network with different speeds for legitimated and delegitimated content. Second, DPI can be used by states in surveillance of citizens.
7. Winner, *The Whale and the Reactor*, 85.
8. Ibid., 95.
9. Galloway, *Protocol*.
10. For a discussion of how software engineering abuses the term "transparency," see Neumann, "Inside Risks"; for an excellent analysis of the engineering of limitations into technology, see Gillespie, *Wired Shut*.
11. Brooks, *The Mythical Man-Month*, 261–264.
12. Ceruzzi, *A History of Modern Computing*, 77.
13. Barbrook and Cameron, "The Californian Ideology"; Turner, *From Counterculture to Cyberculture*.
14. Schaefer and Durham, "On the Social Implications of Invisibility."
15. Woolgar, "Configuring the User."
16. Famiglietti, "Hackers, Cyborgs, and Wikipedians," 108.
17. Bruns, *Blogs, Wikipedia, Second Life, and Beyond*.
18. Turkle, *Life on the Screen*.
19. Enzensberger, "Constituents of a Theory of the Media," 74.
20. Bendrath, "The American Cyber-Angst and the Real World–Any Link"; Dunn Cavelty, *Cyber-security and Threat Politics*.
21. Lawson, "Motivating Cybersecurity."
22. See the IXMaps project at ixmaps.ca for more on Internet Exchange points and their vulnerability to surveillance. See also Klein, *Wiring Up the Big Brother Machine . . . and Fighting It;* Risen and Lichtblau, "Bush Lets U.S. Spy on Callers without Courts."
23. Bamford, "The NSA Is Building the Country's Biggest Spy Center (Watch What You Say)."
24. Best, "Living in the Control Society."
25. See Weber, *The Success of Open Source*, 84–86, for a discussion of copyleft

as the "fundamental" shift of the "optic of intellectual property rights away from protecting the prerogatives of an author toward protecting the prerogatives of generations of users."

26. Fuchs, "The Political Economy of Privacy on Facebook," 141.

27. Maxwell and Miller, *Greening the Media;* "Gartner Says More Than 1 Billion PCs in Use Worldwide and Headed to 2 Billion Units by 2014."

28. Mueller, "Piracy Is Looting—and That's OK."

29. Puckett et al., "Exporting Harm."

30. Maxwell and Miller, *Greening the Media,* 2.

31. For an example of an older computer with a new OS (in this case, Ubuntu Linux), see Brodkin, "Ars Gear." Indeed, much of this book has been written on old computers with Linux Mint. Anyone who wants to switch to Linux, feel free to contact me!

32. boyd and Heer, "Profiles as Conversation."

33. See Bowker, *Memory Practices in the Sciences,* for a discussion of the episteme of "potential memory"—that is, the production of facts out of archives.

34. Deleuze, "Postscript on the Societies of Control."

35. Ibid.

36. "Lorea" is another name for n-1.cc. The name N-1 draws on Deleuze and Guattari's *A Thousand Plateaus;* "Lorea" is a Basque word for flower.

37. Available at https://n-1.cc/pg/expages/read/About/ (accessed June 6, 2012). The original is in Spanish; this is my translation.

38. Crabgrass is part of Riseup.net, the activist networking system that began during the months leading up to the 1999 World Trade Organization meeting in Seattle. Riseup began to offer activists private e-mail services.

39. Sparrow, "Pitfalls of Building Social Media Alternatives (Debate)."

40. Ibid. (roughly at the 6:40 mark).

41. An immediate parallel to Red Hat Linux comes to mind. Red Hat is a for-profit company selling distributions of Linux.

42. Wiese, "On Diaspora's Social Network, You Own Your Data."

43. Dibbell, "The Shadow Web," 63.

44. Akyildiz and Wang, "A Survey on Wireless Mesh Networks," S23.

45. Winner, *The Whale and the Reactor,* chap. 5.

46. Sparrow, "Pitfalls of Building Social Media Alternatives (Debate)" (roughly the 5:50 mark).

47. Mahdian et al., "MyZone."

48. In May 2012, the TalkOpen Terms of Use (previously available at http://talkopen.info/terms) in their entirety were as follows: "By entering this site you take full responsibility for whatever you say or do and acknowledge that TalkOpen.info neither in whole or part is responsible for your actions. With that being said, TalkOpen will NOT sell your information to third parties or give up anyones information to law enforcement unless it is in regards to murder or cp [child pornography]. You have our word. Take it or leave it. :)" In personal correspondence with the author, TalkOpen's founder, XCpherX, promised that if a member were to leave or delete material from TalkOpen, that material would

be permanently deleted. If this was the case, it was in stark contrast to Facebook and Twitter.

49. And this centralization has proved to be a liability in another way: on June 1, 2012, TalkOpen was shut down, its home page replaced by a "Suspended Account" page. It's not clear as of this writing whether it was hacked or suspended by the hosting company.

50. Zurker's books are available for viewing at http://www.zurker.eu/z.r?a= books (accessed June 1, 2012).

51. Mahdian et al., "MyZone," 4.

52. Comment available at http://talkopen.info/view/post:4115#comments (accessed May 31, 2012.) Note that this is no longer available.

53. Dingledine, Mathewson, and Syverson, Tor, 4.

54. Spideralex, "Pitfalls of Building Social Media Alternatives (Debate)."

55. Sparrow, "Pitfalls of Building Social Media Alternatives (Debate)."

56. Kleiner, The Telekommunist Manifesto, 12.

57. Ibid.

Bibliography

Abbate, Janet. *Inventing the Internet*. Cambridge, MA: MIT Press, 1999.

Adams, Samantha. "Sourcing the Crowd for Health Services Improvement: The Reflexive Patient and 'Share-Your-Experience' Websites." *Social Science and Medicine* 72, no. 7 (April 2011): 1069–1076. doi:10.1016/j.socscimed.2011.02.001.

Adorno, Theodor, and Max Horkheimer. "The Culture Industry: Enlightenment as Mass Deception," September 19, 2006. Available at http://www.marxists.org/reference/archive/adorno/1944/culture-industry.htm.

Agger, Ben. *Speeding Up Fast Capitalism: Internet Culture, Work, Families, Food, Bodies*. Boulder, CO: Paradigm Publishers, 2004.

Akin, David. "Casting the Net: Internet Advertising Is Still in Its Infancy, but Many Say It's on the Verge of Phenomenal Growth." *Hamilton Spectator (Ontario, Canada)*, February 24, 1998, FINAL edition.

Akyildiz, I. F., and X. Wang. "A Survey on Wireless Mesh Networks." *Communications Magazine, IEEE* 43, no. 9 (2005): S23–S30.

Albrechtslund, Anders. "Online Social Networking as Participatory Surveillance." *First Monday* 13, no. 3 (March 3, 2008). Available at http://firstmonday.org/article/view/2142/1949.

Aldridge, M. "Confessional Culture, Masculinity and Emotional Work." *Journalism* 2, no. 1 (2001): 91.

Alexie, Sherman. *The Lone Ranger and Tonto Fistfight in Heaven*. Vol. 1. New York: HarperPerennial, 1994.

Alperovitz, Gar. *America beyond Capitalism: Reclaiming Our Wealth, Our Liberty, and Our Democracy*. 2nd ed. Hoboken, NJ: Democracy Collaborative, 2011.

Anderson, Chris. *The Long Tail: Why the Future of Business Is Selling Less of More*. 1st ed. New York: Hyperion, 2006.

Andrejevic, Mark. *iSpy: Surveillance and Power in the Interactive Era*. Lawrence: University Press of Kansas, 2007.

———. *Reality TV: The Work of Being Watched*. Lanham, MD: Rowman and Littlefield Publishers, 2003.

———. "Surveillance and Alienation in the Online Economy." *Surveillance and Society* 8, no. 3 (2011): 278–287.

Arola, Kristin L. "The Design of Web 2.0: The Rise of the Template, the Fall of Design." *Computers and Composition* 27, no. 1 (March 2010): 4–14.

Ashby, W. R. *Design for a Brain*. London: Chapman and Hall, 1960.

Aspray, William. *John von Neumann and the Origins of Modern Computing*. Cambridge, MA: MIT Press, 1990.

Aspray, William, and Martin Campbell-Kelly. *Computer: A History of the Information Machine*. New York: Basic Books, 1997.

Babbage, C. *On the Economy of Machinery and Manufactures*. Kelly, 1832.

Backus, J. "Can Programming Be Liberated from the von Neumann Style?" *Communications of the ACM* 21, no. 8 (August 1978): 613–641.

Bamford, James. "The NSA Is Building the Country's Biggest Spy Center (Watch What You Say)." *Wired*, March 15, 2012. Available at http://www.wired.com/threatlevel/2012/03/ff_nsadatacenter/?utm_source=twitter&utm_medium=socialmedia&utm_campaign=twitterclickthru.

Barbrook, R., and A. Cameron. "The Californian Ideology." *Science as Culture* 6, no. 1 (1996): 44–72.

Barlow, John Perry. "A Declaration of the Independence of Cyberspace." *Electronic Frontier Foundation*, February 8, 1996. Available at https://projects.eff.org /~barlow/Declaration-Final.html.

Barr, J., and L. F. Cabrera. "AI Gets a Brain." *Queue* 4, no. 4 (2006): 29.

Beck, Ulrich. *Risk Society: Towards a New Modernity*. London: Sage Publications, 1992.

Bendrath, R. "The American Cyber-Angst and the Real World–Any Link." In *Bombs and Bandwidth: The Emerging Relationship between Information Technology and Security*, edited by R. Lantham, 49–73. New York: Free Press, 2003.

Benkler, Yochai. "Coase's Penguin, or, Linux and 'The Nature of the Firm.'" *Yale Law Journal* 112, no. 3 (December 2002): 369–446.

———. *The Wealth of Networks: How Social Production Transforms Markets and Freedom*. New Haven, CT: Yale University Press, 2006.

Benkler, Yochai, and H. Nissenbaum. "Commons-Based Peer Production and Virtue." *Journal of Political Philosophy* 14, no. 4 (2006): 394.

Bennett, Colin J. *The Privacy Advocates: Resisting the Spread of Surveillance*. Cambridge, MA: MIT Press, 2010.

Bennett, W. Lance. "New Media Power: The Internet and Global Activism." In *Contesting Media Power: Alternative Media in a Networked World*. Lanham, MD: Rowman and Littlefield, 2003.

Benninger, C. "AJAX Storage: A Look at Flash Cookies and Internet Explorer Persistence." *Foundstone Professional Services and Education, McAfee* (2006).

Benway, J. P. "Banner Blindness: The Irony of Attention Grabbing on the World

Wide Web." *Human Factors and Ergonomics Society Annual Meeting Proceedings* 42 (1998): 463–467.

Benway, J. P., and D. M. Lane. "Banner Blindness: Web Searchers Often Miss 'Obvious' Links." *Internetworking, ITG Newsletter* (1998).

Bercovici, Jeff. "MySpace Users: Fewer Than We Thought?" Financial News. *Daily Finance,* March 3, 2010. Available at http://www.dailyfinance.com/story/company-news/myspaces-dwindling-traffic-looks-even-worse-from-the-inside/19380431/.

Berman, Ari. "Iran's Twitter Revolution." *The Nation,* June 15, 2009. Available at http://www.thenation.com/blogs/notion/443634.

Bernard, M. L. "Developing Schemas for the Location of Common Web Objects." *Human Factors and Ergonomics Society Annual Meeting Proceedings* 45 (2001): 1161–1165.

Berners-Lee, Tim, and Mark Fischetti. *Weaving the Web: The Original Design and Ultimate Destiny of the World Wide Web by Its Inventor.* New York: Harper-Collins, 2000.

Berry, David M. "The Poverty of Networks." *Theory, Culture and Society* 25, no. 7–8 (2008): 364–372.

———. "The Uses of Object-Oriented Ontology." *Stunlaw,* May 2012. Available at http://stunlaw.blogspot.nl/2012/05/uses-of-object-oriented-ontology.html.

Best, Kirsty. "Living in the Control Society: Surveillance, Users and Digital Screen Technologies." *International Journal of Cultural Studies* 13, no. 1 (January 1, 2010): 5–24. doi:10.1177/1367877909348536.

Bianco, Jamie Skye. "Social Networking and Cloud Computing: Precarious Affordances for the 'Prosumer.'" *WSQ: Women's Studies Quarterly* 37, no. 1 (2009): 303–312. doi:10.1353/wsq.0.0146.

Bijker, Wiebe E., Thomas Parke Hughes, and T. J. Pinch. *The Social Construction of Technological Systems: New Directions in the Sociology and History of Technology.* Cambridge, MA: MIT Press, 1987.

Bogost, Ian. *Alien Phenomenology, or What It's Like to Be a Thing.* Minneapolis: University of Minnesota Press, 2012.

Boh, W. F, C. Soh, and S. Yeo. "Standards Development and Diffusion: A Case Study of RosettaNet." *Communications of the ACM* 50, no. 12 (2007): 57–62.

Boorstin, Julia. "MySpace Finally Sold for Some $35 Million." *Christian Science Monitor,* June 29, 2011. Available at http://www.csmonitor.com/Business/Latest-News-Wires/2011/0629/MySpace-finally-sold-for-some-35-million.

Bosch, Jan. *Design and Use of Software Architectures: Adopting and Evolving a Product-Line Approach.* Reading, MA: Addison-Wesley, 2000.

Boshmaf, Y., I. Muslukhov, K. Beznosov, and M. Ripeanu. "The Socialbot Network: When Bots Socialize for Fame and Money." In *Proceedings of the 27th Annual Computer Security Applications Conference,* 93–102. New York: ACM, 2011.

Bottero, W. "Class Identities and the Identity of Class." *Sociology* 38, no. 5 (2004): 985–1003.

Bowker, Geoffrey C. "How to Be Universal: Some Cybernetic Strategies, 1943–70." *Social Studies of Science* 23, no. 1 (1993): 107–127.

———. *Memory Practices in the Sciences.* Inside Technology. Cambridge, MA: MIT Press, 2005.

boyd, danah. "Facebook and MySpace Users Are Clearly Divided along Class Lines." *AlterNet,* September 3, 2009. Available at http://www.alternet.org/media/142356/facebook_and_myspace_users_are_clearly_divided_along_class_lines?page=entire.

———. "Facebook and 'Radical Transparency' (A Rant)." *Apophenia,* May 14, 2010. Available at http://www.zephoria.org/thoughts/archives/2010/05/14/facebook-and-radical-transparency-a-rant.html?utm_source=feedburner&utm_medium=feed&utm_campaign=Feed%3A+zephoria%2Fthoughts+%28apophenia%29.

———. "Facebook's Privacy Trainwreck: Exposure, Invasion, and Social Convergence." *Convergence* 14, no. 1 (2008): 13–20.

———. "Friendster and Publicly Articulated Social Networking." *Conference on Human Factors and Computing Systems,* 1–5, 2004.

———. "Viewing American Class Divisions through Facebook and MySpace," June 24, 2007. Available at http://www.danah.org/papers/essays/Class Divisions.html.

boyd, danah, and N. B. Ellison. "Social Network Sites: Definition, History, and Scholarship." *Journal of Computer-Mediated Communication* 13, no. 1 (2008): 210–230.

boyd, danah, and Jeffrey Heer. "Profiles as Conversation: Networked Identity Performance on Friendster." In *Persistent Conversation Track,* 1–10. Kauai, HI: IEEE Computer Society, 2006.

boyd, danah, and Henry Jenkins. "MySpace and Deleting Online Predators Act (DOPA)." *MIT Tech Talk,* May 26, 2006. Available at http://www.danah.org/papers/MySpaceDOPA.html.

Bratich, Jack. "'Nothing Is Left Alone for Too Long': Reality Programming and Control Society Subjects." *Journal of Communication Inquiry* 30, no. 1 (2006): 65–83.

———. "User-Generated Discontent." *Cultural Studies* 25 (September 2011): 621–640. doi:10.1080/09502386.2011.600552.

Braverman, Harry. *Labor and Monopoly Capital: The Degradation of Work in the Twentieth Century.* New York: Monthly Review Press, 1975.

Brenner, Robert. *The Boom and the Bubble: The US in the World Economy.* London: Verso, 2002.

Brodkin, Jon. "Ars Gear: The Old Windows PC Resurrected with Ubuntu." *Ars Technica,* May 27, 2012. Available at http://arstechnica.com/gadgets/2012/05/ars-gear-the-old-windows-pc-resurrected-with-ubuntu/.

Brooks, Frederick P. *The Mythical Man-Month: Essays on Software Engineering, Anniversary Edition.* 2nd ed. Reading, MA: Addison-Wesley Professional, 1995.

Bruner, Rick E. *The Decade in Online Advertising: 1994–2004.* DoubleClick, April 2005. Available at http://www.google.com/doubleclick/pdfs/DoubleClick-04-2005-The-Decade-in-Online-Advertising.pdf.

Bruns, Axel. *Blogs, Wikipedia, Second Life, and Beyond: From Production to Produsage.* New York: Peter Lang, 2008.

————. "The Future Is User-Led: The Path towards Widespread Produsage." *Fibreculture* no. 11 (2008). Available at http://journal.fibreculture.org/issue11/issue11_bruns.html.

Burawoy, M. "Reflections on the Class Consciousness of Hungarian Steelworkers." *Politics and Society* 17, no. 1 (1989): 1–34.

Burke, M., N. Gorman, E. Nilsen, and A. Hornof. "Banner Ads Hinder Visual Search and Are Forgotten." *CHI'04 Extended Abstracts on Human Factors in Computing Systems*, 1139–1142, 2004.

Cantoni, V., and S. Levialdi. "Matching the Task to an Image Processing Architecture (for Overcoming Von Neumann Bottleneck)." *Computer Vision, Graphics, and Image Processing* 22 (1983): 301–309.

Cashmore, Pete. "MySpace, America's Number One." *Mashable*, July 11, 2006. Available at http://mashable.com/2006/07/11/myspace-americas-number-one/.

Center for Digital Democracy, U.S. Public Interest Research Group, and World Privacy Forum. *Complaint, Request for Investigation, Injunction, and Other Relief: Google, Yahoo, PubMatic, TARGUSinfo, MediaMath, eXelate, Rubicon Project, AppNexus, Rocket Fuel, and Others Named Below.* Complaint to Federal Trade Commission. Washington, DC: Center for Digital Democracy, April 5, 2010. Available at http://www.centerfordigitaldemocracy.org/sites/default/files/20100407-FTCfiling.pdf.

Ceruzzi, Paul E. *A History of Modern Computing.* Cambridge, MA: MIT Press, 2003.

Chang, Briankle G. "To the Archive: A Postal Tale." *Communication and Critical/Cultural Studies* 7, no. 2 (2010): 202. doi:10.1080/14791421003790872.

Chatterjee, P., D. L. Hoffman, and T. P. Novak. "Modeling the Clickstream: Implications for Web-Based Advertising Efforts." *Marketing Science* (2003): 520–541.

Chikofsky, E. J. and, J. H. Cross. "Reverse Engineering and Design Recovery: A Taxonomy." *Software, IEEE* 7, no. 1 (1990): 13–17.

Christ, Julian P., and André Slowak. "Why Blu-Ray vs. HD-DVD Is Not VHS vs. Betamax: The Co-evolution of Standard-Setting Consortia." *SSRN eLibrary* (April 28, 2009). Available at http://papers.ssrn.com/sol3/papers.cfm?abstract_id=1626224.

Christian, Jean. "Real Vlogs: The Rules and Meanings of Online Personal Videos." *First Monday* 14, no. 11 (November 2, 2009). Available at http://www.uic.edu/htbin/cgiwrap/bin/ojs/index.php/fm/article/view/2699/2353.

Clarke, Roger. "The Digital Persona and Its Application to Data Surveillance." *The Information Society* 10, no. 2 (1994): 77–92. doi:10.1080/01972243.1994.9960160.

Cleaver, H. M., Jr. "The Zapatista Effect: The Internet and the Rise of an Alternative Political Fabric." *Journal of International Affairs* 51, no. 2 (1998): 621–622.

Cleland, Kim. "Media Buying and Planning: Marketers Want Solid Data on Value of Internet Ad Buys: Demand Swells for Information That Compares Media Options." *Advertising Age*, August 3, 1998.

Coast Artillery Field Manual. U.S. G.P.O., 1940.

Cohen, Noam. "An Ad Blocker Opens the Gate, Ever So Slightly." *New York Times,* January 1, 2012, sec. Business Day / Media and Advertising. Available at https://www.nytimes.com/2012/01/02/business/media/adblock-plus-allow ing-some-online-advertisements.html.

————. "In Google-Verizon Deal, Fears for Privacy." *New York Times,* August 15, 2010, sec. Business Day / Media and Advertising. Available at http://www.nytimes.com/2010/08/16/business/media/16link.html?ref=googleinc.

Colburn, Timothy, and Gary Shute. "Abstraction in Computer Science." *Minds and Machines* 17, no. 2 (July 1, 2007): 169–184. doi:10.1007/s11023-007-9061-7.

Collins, Glenn. "Trade Groups Propose Web Banner Guidelines." *New York Times,* December 12, 1996, Late Edition—Final edition.

Considine, Austin. "Twitter Followers for Sale." *New York Times,* August 22, 2012, sec. Fashion and Style. Available at https://www.nytimes.com/2012/08/23/fashion/twitter-followers-for-sale.html.

Constine, Jason. "Myspace Lost $43M in 2012: Here's Why Launching a Profit-Thin Music Service Could Nail the Coffin Shut." *TechCrunch,* November 19, 2012. Available at http://techcrunch.com/2012/11/19/myspace-music-stream ing-service-no/.

Consumer Data Privacy in a Networked World: A Framework for Protecting Privacy and Promoting Innovation in the Global Digital Economy. Washington, DC: The White House, February 23, 2012.

Costolo, Richard. "Twitter, Inc Registration Statement on Form S-1," October 3, 2013. Available at http://www.sec.gov/Archives/edgar/data/1418091/000119 312513390321/d564001ds1.htm.

Coté, Mark, and Jennifer Pybus. "Learning to Immaterial Labour 2.0: MySpace and Social Networks." *Ephemera* 7, no. 1 (2007): 88–106.

Couldry, Nick. *Media Rituals: A Critical Approach.* London: Routledge, 2003.

Creegan, Jennifer. "Dodge Becomes First Advertiser to Tap Microsoft's IAB Award-Winning Filmstrip." *Microsoft Advertising,* May 13, 2011. Available at http://community.microsoftadvertising.com/blogs/advertising/archive/2011/05/13/dodge-microsoft-advertising-iab-filmstrip.aspx.

Cuban, Mark. "An Open Facebook API Vs Google OpenSocial." *Blog Maverick,* November 4, 2007. Available at http://blogmaverick.com/2007/11/04/an-open-facebook-api-vs-google-opensocial/.

Davis, Martin. "Mathematical Logic and the Origin of Modern Computers." In *The Universal Turing Machine: A Half-Century Survey,* edited by Rolf Herken, 149–174. Oxford, UK: Oxford University Press, 1988.

Davis, Michael. "Will Software Engineering Ever Be Engineering?" *Communications of the ACM* 54, no. 11 (November 1, 2011): 32–34. doi:10.1145/2018396.2018407.

Daw, David. "10 Twitter Bot Services to Simplify Your Life." *PC World,* October 23, 2011. Available at https://www.pcworld.com/article/242338/10_twitter_bot_services_to_simplify_your_life.html.

Dean, Jodi. "The Networked Empire: Communicative Capitalism and the Hope for Politics." In *Empire's New Clothes: Reading Hardt and Negri*, edited by P. A Passavant and J. Dean, 265—288. New York: Psychology Press, 2004.

DeBenedictis, E. P., and S. C. Johnson. "Extending Unix for Scalable Computing." *IEEE Computer* 26, no. 11 (1993): 43–53.

Deleuze, Gilles. "Postscript on the Societies of Control." *October* 59 (Winter 1992): 3–7. doi:10.2307/778828.

DeLuca, Kevin Michael. *Image Politics: The New Rhetoric of Environmental Activism*. New York: Lawrence Erlbaum, 2005.

Derrida, Jacques. *Archive Fever: A Freudian Impression*. Religion and Postmodernism. Chicago: University of Chicago Press, 1996.

Desilva and Phillips, LLC. *Online Ad Networks: Monetizing the Long Tail*. New York: Desilva and Phillips, LLC, 2008.

De Vries, Lloyd. "MySpace Tightens Age Restrictions." News. *CBS News,* June 21, 2006. Available at http://www.cbsnews.com/stories/2006/06/21/tech/main1736549.shtml.

Dibbell, Julian. "The Shadow Web." *Scientific American* 306, no. 3 (February 14, 2012): 60–65. doi:10.1038/scientificamerican0312-60.

Dickinson, A. *An Optical Respite from the Von Neumann Bottleneck*. Holmdel, NJ: AT&T Bell Labs, 1992.

Dingledine, Roger, Nick Mathewson, and Paul Syverson. *Tor: The Second-Generation Onion Router,* 2004. Available at http://stinet.dtic.mil/oai/oai?&verb=getRecord&metadataPrefix=html&identifier=ADA465464.

DiNucci, Darcy. *Elements of Web Design*. Berkeley, CA: Peachpit Press, 1997.

Douglas, Susan J. *Inventing American Broadcasting, 1899–1922*. Johns Hopkins Studies in the History of Technology. Baltimore, MD: Johns Hopkins University Press, 1987.

Dreze, X., and F. Zufryden. "Is Internet Advertising Ready for Prime Time?" *Journal of Advertising Research* 38 (1998): 7–18.

Dugdale, Anni. "Materiality: Juggling Sameness and Difference." In *Actor Network Theory and After,* edited by John Hassard and John Law, 113–135. Oxford, UK: Blackwell/Sociological Review, 1999.

Dunn Cavelty, Myriam. *Cyber-security and Threat Politics: US Efforts to Secure the Information Age*. Abingdon, UK: Routledge, 2007.

Dyer-Witheford, Nick. *Cyber-Marx: Cycles and Circuits of Struggle in High-Technology Capitalism*. Urbana: University of Illinois Press, 1999.

Ebensperger, Lukas, Suparna Choudhury, and Jan Slaby. "Designing the Lifeworld: Selfhood and Architecture from a Critical Neuroscience Perspective." In *Cognitive Architecture: From Bio-politics to Noo-politics; Architecture and Mind in the Age of Communication and Information,* edited by Deborah Hauptman and Warren Neidich, 232–246. Rotterdam, Netherlands: 010 Publishers, 2010.

Ebersman, David A. "Facebook, Inc. Registration Statement on Form S-1," February 1, 2012. Available at http://www.sec.gov/Archives/edgar/data/1326801/000119312512034517/d287954ds1.htm.

Eckersley, Peter. "How Unique Is Your Web Browser?" In *Privacy Enhancing Technologies,* edited by Mikhail J. Atallah and Nicholas J. Hopper, 6205:1–18. Berlin: Springer, 2010. Available at http://www.springerlink.com/index/10.1007/978-3-642-14527-8_1.

Eckert, J. P. "Disclosure of a Magnetic Calculating Machine." *Unpublished Typescript* (1945): 525–539.

Edwards, Paul N. *The Closed World: Computers and the Politics of Discourse in Cold War America.* Inside Technology. Cambridge, MA: MIT Press, 1996.

Egyedi, T. M. "Consortium Problem Redefined: Negotiating 'Democracy' in the Actor Network on Standardization." *International Journal of IT Standards and Standardization Research* 1, no. 2 (2002): 22–38.

Elowitz, Ben. "Traditional Ways of Judging 'Quality' in Published Content Are Now Useless." *paidContent,* May 3, 2010. Available at http://paidcontent.org/article/419-traditional-ways-of-judging-quality-in-published-content-are-now-useless/.

"Enciclopedia Discusión: Por Qué Estamos Aquí y No en Es.wikipedia.org." *La Enciclopedia Libre Universal en Español,* October 26, 2006. Available at http://enciclopedia.us.es/index.php/Enciclopedia_Discusi%C3%B3n:Por_qu%C3%A9_estamos_aqu%C3%AD_y_no_en_es.wikipedia.org.

Enyedy, Edgar. "Five Messages." *OSDir,* February 19, 2002. Available at http://osdir.com/ml/science.linguistics.wikipedia.international/2002-02/msg00053.html.

———. "Good Luck with Your wikiPAIDia." *OSDir,* February 17, 2002. Available at http://osdir.com/ml/science.linguistics.wikipedia.international/2002-02/msg00038.html.

———. "Spanish Wikipedia." *OSDir,* January 22, 2002. Available at http://osdir.com/ml/science.linguistics.wikipedia.international/2002-01/msg00116.html.

Enyedy, Edgar, and Nathaniel Tkacz. "'Good Luck with Your WikiPAIDia': Reflections on the 2002 Fork of the Spanish Wikipedia." *Network Cultures,* January 2011. Available at http://networkcultures.org/wpmu/cpov/lang/de/2011/01/15/spanish_fork/.

Enzensberger, H. M. "Constituents of a Theory of the Media." In *Media Studies: A Reader,* edited by Paul Marris and Sue Thornham, 68–91. 2nd ed. New York: New York University Press, 2000.

"The Face behind Facebook." *60 Minutes,* January 13, 2008. Available at http://www.cbsnews.com/video/watch/?id=3706601n&tag=related;photovideo.

"Facebook: Largest, Fastest Growing Social Network." *TechTree.com India,* August 13, 2008. Available at http://www.techtree.com/India/News/Facebook_Largest_Fastest_Growing_Social_Network/551-92134-643.html.

"Facebook Revenue Surges 60% on Strong Ad Sales." *BBC News,* October 31, 2013. Available at http://www.bbc.co.uk/news/business-24751441.

Famiglietti, Andrew A. "Hackers, Cyborgs, and Wikipedians: The Political Economy and Cultural History of Wikipedia." Bowling Green, OH: Bowling Green State University, 2011. Available at http://rave.ohiolink.edu/etdc/view?acc_num=bgsu1300717552.

Federal Trade Commission. *Ehavorial Advertising: Tracking, Targeting, and Technology.* Washington, DC: Federal Trade Commission, November 1, 2007. Available at http://www.ftc.gov/bcp/workshops/ehavioral/index.shtml.

———. *Self-Regulatory Principles for Online Behavioral Advertising.* Washington DC: Federal Trade Commission, February 2009. Available at http://www.ftc .gov/os/2009/02/P085400bchavadrcport.pdf.

Finelli, Roberto. *Astrazione e Dialettica Dal Romanticismo Al Capitalismo: (saggio Su Marx).* Rome: Biblioteca Di Cultura (Bulzoni Editore), 1987.

Fisher, Eran. *Media and New Capitalism in the Digital Age: The Spirit of Networks.* New York: Palgrave Macmillan, 2010.

Fixmer, Andy, and Sarah Rabil. "MySpace Sale, Merger or Spinoff Being Weighed by News Corp., Official Says." *Bloomberg,* January 12, 2011. Available at http://www.bloomberg.com/news/2011-01-12/myspace-sale-merger-or-spin off-being-weighed-by-news-corp-official-says.html.

Flanagin, Andrew J., Craig Flanagin, and Jon Flanagin. "Technical Code and the Social Construction of the Internet." *New Media and Society* 12, no. 2 (March 1, 2010): 179–196. doi:10.1177/1461444809341391.

Foresman, Chris. "Myspace Concedes to Facebook, Changes Focus." *Ars Technica,* December 2010. Available at http://arstechnica.com/web/news/2010/10/ myspace-concedes-to-facebook-changes-focus.ars.

Foucault, Michel. *The Archaeology of Knowledge; and The Discourse on Language.* New York: Pantheon Books, 1972.

———. *Discipline and Punish.* New York: Vintage, 1979.

———. *The Order of Things: An Archaeology of the Human Sciences.* London: Tavistock Publications, 1970.

———. "The Subject and Power." In *The Essential Foucault: Selections from Essential Works of Foucault, 1954–1984,* edited by Paul Rabinow and Nikolas S. Rose, 126–144. New York: New Press, 2003.

Froehling, O. "The Cyberspace 'War of Ink and Internet' in Chiapas, Mexico." *Geographical Review* 87, no. 2 (1997): 291–307.

Fuchs, Christian. "The Political Economy of Privacy on Facebook." *Television and New Media* 13, no. 2 (March 1, 2012): 139–159. doi:10.1177/1527476411415699.

Fuglsang, Martin, and Bent Meier Sørensen. *Deleuze and the Social.* Edinburgh, Scotland: Edinburgh University Press, 2006. Available at http://site.ebrary. com/id/10131990.

Fuller, Matthew. *Behind the Blip: Essays on the Culture of Software.* New York: Autonomedia, 2003.

Fuller, Matthew, and Andrew Goffey. "Digital Infrastructures and the Machinery of Topological Abstraction." *Theory, Culture and Society* 29, no. 4–5 (July 1, 2012): 311–333. doi:10.1177/0263276412450466.

Galison, Peter. "The Ontology of the Enemy: Norbert Wiener and the Cybernetic Vision." *Critical Inquiry* 21 (Autumn 1994): 228–266.

Galloway, Alexander. *Protocol: How Control Exists after Decentralization.* Cambridge, MA: MIT Press, 2004.

Galloway, Alexander, and Eugene Thacker. *The Exploit: A Theory of Networks.* Minneapolis: University of Minnesota Press, 2007.

"Gartner Says More Than 1 Billion PCs In Use Worldwide and Headed to 2 Billion Units by 2014." *Gartner,* June 23, 2008. Available at https://www.gartner.com/it/page.jsp?id=703807.

Gates, Kelly A. *Our Biometric Future: Facial Recognition Technology and the Culture of Surveillance.* New York: New York University Press, 2011.

Gehl, Robert W. "Distributed Centralization: Web 2.0 as a Portal into Users' Lives." *Lateral* 1, no. 1 (March 2012). Available at http://lateral.culturalstudiesassociation.org/issue1/content/gehl.html.

———. "Ladders, Samurai, and Blue Collars: Personal Branding in Web 2.0." *First Monday* 16, no. 9 (September 5, 2011).

———. "'Why I Left Facebook': Stubbornly Refusing to Not Exist Even after Opting Out of Mark Zuckerberg's Social Graph." In *Unlike Us Reader: Social Media Monopolies and Their Alternatives,* edited by Geert Lovink and Miriam Rausch, 220–238. Amsterdam: Institute of Network Cultures, 2013.

———. "YouTube as Archive: Who Will Curate This Digital Wunderkammer?" *International Journal of Cultural Studies* 12, no. 1 (2009): 43–60.

Gehl, Robert W., and Sarah Bell. "Heterogeneous Software Engineering: Garmisch 1968, Microsoft Vista, and a Methodology for Software Studies." *Computational Culture* no. 2 (2012). Available at http://computationalculture.net/article/heterogeneous-software-engineering-garmisch-1968-microsoft-vista-and-a-methodology-for-software-studies.

Gerben, Chris. "Privileging the 'New' in New Media Literacy: The Future of Democracy, Economy, and Identity in 21st Century Texts." Presented at Media in Transition 5, Cambridge, MA, 2009. Available at http://web.mit.edu/comm-forum/mit6/papers/Gerben.pdf

Giddens, Anthony. *Modernity and Self-Identity: Self and Society in the Late Modern Age.* Stanford, CA: Stanford University Press, 1991. Available at http://www.loc.gov/catdir/description/cam024/91065170.html and http://www.loc.gov/catdir/toc/cam026/91065170.html.

Gillespie, Tarleton. "Designed to 'Effectively Frustrate': Copyright, Technology and the Agency of Users." *New Media Society* 8, no. 4 (August 1, 2006): 651–669. doi:10.1177/1461444806065662.

———. *Wired Shut.* Cambridge, MA: MIT Press, 2007.

Glott, R., P. Schmidt, and R. Ghosh. "Wikipedia Survey—Overview of Results." *United Nations University: Collaborative Creativity Group.* Available at http://www.Wikipediasurvey.org/docs/Wikipedia_Overview_15March2010-FINAL.Pdf (November 29, 2010).

Goffey, Andrew. "Algorithm." In *Software Studies a Lexicon,* edited by Matthew Fuller, 15–20. Cambridge, MA: MIT Press, 2008.

Goldman, David. "MySpace to Cut 30%, or 430 Jobs." *CNNMoney,* June 16, 2009. Available at http://money.cnn.com/2009/06/16/technology/myspace_layoffs/index.htm.

Golumbia, David. *The Cultural Logic of Computation.* Cambridge, MA: Harvard University Press, 2009.

Governor, James, Dion Hinchcliffe, and Duane Nickull. *Web 2.0 Architectures.* 1st ed. Sebastopol, CA: O'Reilly Media, 2009.

Granka, Laura, Helene Hembrooke, and Geri Gay. "Location Location Location: Viewing Patterns on WWW Pages." In *Proceedings of the 2006 Symposium on Eye Tracking Research and Applications*, 43–43. ETRA '06. New York: ACM, 2006. doi:10.1145/1117309.1117328.

Grier, David Alan. "Gertrude Blanch of the Mathematical Tables Project." *IEEE Annals of the History of Computing* 19, no. 4 (December 1997): 18–27. doi:10 .1109/85.627896.

———. *When Computers Were Human*. Princeton, NJ: Princeton University Press, 2005.

Guide to the Software Engineering Body of Knowledge (SWEBOK). Los Alamitos, CA: IEEE Computer Society, 2004. Available at http://www.computer.org/ portal/web/swebok/htmlformat.

Hall, S. "Gramsci's Relevance for the Study of Race and Ethnicity." *Stuart Hall: Critical Dialogues in Cultural Studies* (1996): 5–27.

Haraway, Donna Jeanne. "A Manifesto for Cyborgs: Science, Technology, and Socialist Feminism in the 1980s." In *The Haraway Reader*, 7–45. New York: Routledge, 2004.

Hardt, Michael, and Antonio Negri. *Empire*. Cambridge, MA: Harvard University Press, 2000.

———. *Multitude: War and Democracy in the Age of Empire*. New York: Penguin Press, 2004.

Hartenstein, R. "Data-Stream-Based Computing: Models and Architectural Resources." *INFORMACIJE MIDEM-LJUBLJANA* 33, no. 4 (2003): 228–235.

———. "The Digital Divide of Computing." *Proceedings of the 1st Conference on Computing Frontiers*, 357–362, 2004.

Hawker, Mark D. *The Developer's Guide to Social Programming: Building Social Context Using Facebook, Google Friend Connect, and the Twitter API*. Upper Saddle River, NJ: Addison Wesley, 2011.

Hawkins, R. "The Rise of Consortia in the Information and Communication Technology Industries: Emerging Implications for Policy." *Telecommunications Policy* 23, no. 2 (1999): 159–173.

Hochschild, Arlie Russell. *The Managed Heart: Commercialization of Human Feeling*. Berkeley: University of California Press, 1983.

Hodges, Andrew. "Alan Turing and the Turing Machine." In *The Universal Turing Machine: A Half-Century Survey*, edited by Rolf Herken, 3–15. Oxford, UK: Oxford University Press, 1988.

Holdener, Anthony T. *Ajax the Definitive Guide*. Farnham, UK: O'Reilly, 2008.

Howard, Alex. "2011 Trends: National Strategy for Trusted Identities in Cyberspace Highlights Key Online Privacy, Security Challenges." *Gov20.govfresh*, January 7, 2011. Available at http://gov20.govfresh.com/2011-trends-national-strategy-for-trusted-identities-in-cyberspace-highlights-key-online-privacy-security-challenges/.

"How Digg Uncovers the News." *BusinessWeek*, November 21, 2005. Available at http://www.businessweek.com/magazine/content/05_47/b3960426.htm.

Howe, Jeff. *Crowdsourcing*. New York: Random House, 2008.

———. "The Rise of Crowdsourcing." *Wired Magazine* 14, no. 6 (2006): 1–4.

Huws, Ursula. *The Making of a Cybertariat: Virtual Work in a Real World.* New York: Monthly Review Press, 2003.
Hwang, Tim, Ian Pearce, and Max Nanis. "Socialbots: Voices from the Fronts." *Interactions* 19, no. 2 (April 2012): 38–45.
"IAB Misstep." *Advertising Age,* July 7, 1996.
IAB Pop-up Task Force. "Pop-Up Guidelines and Best Practices: A Discussion around Our Final Recommendation." September 2004. Available at http://www.iab.net/media/file/Pop-UpGuidelinesIndustryReview-Sept04.pdf.
IAB User-Generated Content and Social Media Committee. *Social Advertising Best Practices,* May 2009. Available at http://www.iab.net/media/file/Social-Advertising-Best-Practices-0509.pdf.
———. *Social Media Ad Metrics Definitions.* Interactive Advertising Bureau, May 2009.
Illouz, Eva. *Cold Intimacies: The Making of Emotional Capitalism.* Cambridge, UK: Polity, 2007.
Ingle, Kathryn A. *Reverse Engineering.* New York: McGraw-Hill, 1994.
Interactive Advertising Bureau. "Ad Unit Guidelines." *Interactive Advertising Bureau,* February 28, 2011. Available at http://www.iab.net/iab_products_and_industry_services/1421/1443/1452.
———. *Glossary of Interactive Advertising Terms V. 2.0.* Interactive Advertising Bureau, n.d.
———. "IAB Announces Winners of 'Rising Stars' Competition: New Brand-Friendly Ad Formats Aim to Spur Greater Creativity in Interactive Advertising," *Interactive Advertising Bureau,* February 28, 2011. Available at http://www.iab.net/about_the_iab/recent_press_releases/press_release_archive/press_release/pr-022811_risingstars.
———. "IAB Issues First-Ever Guidelines for Interactive Audience Measurement and Advertising Campaign Reporting and Audits." *Interactive Advertising Bureau,* January 15, 2002. Available at http://www.iab.net/about_the_iab/recent_press_releases/press_release_archive/press_release/4471.
———. "IAB Publishes Glossary of Terms." *Interactive Advertising Bureau,* October 17, 2001. Available at http://www.iab.net/about_the_iab/recent_press_releases/press_release_archive/press_release/4448.
———. "Rich Media Creative Guidelines." *Interactive Advertising Bureau,* 2008. Available at http://www.iab.net/iab_products_and_industry_services/508676/508767/Rich_Media.
———. "Voluntary Guidelines for Banner Advertising and Process for Exploring Future Internet Advertising Forms Announced by Industry Groups." *Interactive Advertising Bureau,* December 10, 1996. Available at http://www.iab.net/about_the_iab/recent_press_releases/press_release_archive/press_release/4219.
Interactive Advertising Bureau, American Association of Advertising Agencies, Association of National Advertisers, Better Business Bureau, and Direct Marketer's Association. *Self-Regulatory Principles for Online Behavioral Advertising.* Washington, DC: Interactive Advertising Bureau, July 2009.
Interactive Advertising Bureau and American Association of Advertising Agencies.

Standard Terms and Conditions for Internet Advertising for Media Buys One Year or Less (version 3.0). Contract, n.d.

Interactive Advertising Bureau (UK). *IAB Jargon Buster.* Interactive Advertising Bureau (UK), January 2002. Available at http://www.iabuk.net/en/1/jargon-buster.html.

"Iran's Twitter Revolution." *Washington Times,* June 16, 2009. Available at http://www.washingtontimes.com/news/2009/jun/16/irans-twitter-revolution/.

Jochym, Pawel. "Re: [Intlwiki-l] Test the New Software! Test.wikipedia.com Is Up!" *OSDir,* January 19, 2002. Available at http://osdir.com/ml/science.linguistics.wikipedia.international/2002-01/msg00115.html.

Kahn, Richard, and Douglas Kellner. "New Media and Internet Activism: From the 'Battle of Seattle' to Blogging." *New Media and Society* 6, no. 1 (2004): 87–95.

Kay, B. "Extending Tort Liability to Creators of Fake Profiles on Social Networking Websites." *Chicago-Kent Journal of Intellectual Property* (2010): 1–24.

Kelly, Kevin. *New Rules for the New Economy: 10 Radical Strategies for a Connected World.* New York: Penguin, 1999.

Kirschenbaum, Matthew G. *Mechanisms: New Media and the Forensic Imagination.* Cambridge, MA: MIT Press, 2012.

Kitchin, Rob, and Martin Dodge. *Code/Space: Software and Everyday Life.* Cambridge, MA: MIT Press, 2011.

Klapp, O. E. "Meaning Lag in the Information Society." *Journal of Communication* 32, no. 2 (1982): 56–66.

Klein, Mark. *Wiring Up the Big Brother Machine . . . and Fighting It.* Charleston, SC: BookSurge Publishing, 2009.

Kleiner, Dmytri. *The Telekommunist Manifesto.* Network Notebooks 3. Amsterdam: Institute of Network Cultures, 2010.

Knudson, J. W. "Rebellion in Chiapas: Insurrection by Internet and Public Relations." *Media, Culture and Society* 20, no. 3 (1998): 507–518.

Komanduri, Saranga, Richard Shay, Greg Norcie, Blase Ur, and Lorrie Faith Cranor. "AdChoices? Compliance with Online Behavioral Advertising Notice and Choice Requirements." *I/S: A Journal of Law and Policy for the Information Society* 7 (2012): 603–721.

Lampland, Martha, and Susan Leigh Star, eds. *Standards and Their Stories: How Quantifying, Classifying, and Formalizing Practices Shape Everyday Life.* Cornell, NY: Cornell University Press, 2008.

Lanier, Jaron. *You Are Not a Gadget: A Manifesto.* New York: Knopf, 2010.

Latour, Bruno. *Reassembling the Social: An Introduction to Actor-Network-Theory.* Oxford, UK: Oxford University Press, 2005.

Law, John. *Aircraft Stories Decentering the Object in Technoscience.* Durham, NC: Duke University Press, 2002.

———. "Making a Mess with Method." January 19, 2006. Available at http://www.heterogeneities.net/publications/Law2006MakingaMesswithMethod.pdf.

———. "Notes on the Theory of the Actor-Network: Ordering, Strategy, and Heterogeneity." *Systemic Practice and Action Research* 5, no. 4 (August 1992): 379–393. doi:10.1007/BF01059830.

————. *A Sociology of Monsters: Essays on Power, Technology and Domination.* London: Routledge, 1991.

————. "Technology and Heterogeneous Engineering: The Case of Portuguese Expansion." In *The Social Construction of Technological Systems: New Directions in the Sociology and History of Technology,* edited by Wiebe Bijker, Thomas Parke Hughes, and Trevor Pinch, 111–134. Cambridge, MA: MIT Press, 1989.

Lawson, Sean. "Motivating Cybersecurity: Assessing the Status of Critical Infrastructure as an Object of Cyber Threats." In *Securing Critical Infrastructures and Industrial Control Systems: Approaches for Threat Protection,* edited by Atta Badii and Christopher Laing. Hershey, PA: IGI Global, 2012.

Layton, Julia. "How Digg Works." *How Stuff Works,* 2006. Available at http://com puter.howstuffworks.com/internet/social-networking/networks/digg.htm.

Lazzarato, Maurizio. "The Concepts of Life and the Living in the Societies of Control." In *Deleuze and the Social,* edited by M. Fuglsang and B. Sørensen, 171–190. Edinburgh, Scotland: Edinburgh University Press, 2006.

Learmonth, Michael. "Google to Add +1 Social Layer to Display Ads." *Advertising Age,* September 20, 2011. Available at http://adage.com/article/special-report-digital-west/google-add-1-social-layer-display-ads/229892/.

Lee, Gwendolyn K., and Robert E. Cole. "From a Firm-Based to a Community-Based Model of Knowledge Creation: The Case of the Linux Kernel Development," *Organization Science* 14, no. 6 (2003): 633–649.

Leistert, Oliver. "How to Put Up a Facebook Resistance." *Centre for Internet and Society,* February 21, 2012. Available at http://cis-india.org/digital-natives/me dia-coverage/facebook-resistance.

Lessig, Lawrence. *Code: Version 2.0.* New York: Basic, 2006.

Light, Jennifer S. "When Computers Were Women." *Technology and Culture* 40, no. 3 (1999): 455–483.

Lih, Andrew. *The Wikipedia Revolution: How a Bunch of Nobodies Created the World's Greatest Encyclopedia.* New York: Hyperion, 2009.

Linus. "My Views on the International Wikipedias and What They Should Do." *OSDir,* October 5, 2001. Available at http://osdir.com/ml/science.linguistics. wikipedia.international/2001-10/msg00004.html.

Liskov, Barbara, and John Guttag. *Program Development in Java: Abstraction, Specification, and Object-Oriented Design.* Boston: Addison-Wesley, 2000.

Liu, Hugo. "Social Network Profiles as Taste Performances." *Journal of Computer-Mediated Communication* 13, no. 1 (2007). Available at http://jcmc.indiana .edu/vol13/issue1/liu.html.

Liu, Hugo, Pattie Maes, and Glorianna Davenport. "Unraveling the Taste Fabric of Social Networks." *International Journal on Semantic Web and Information Systems* 2, no. 1 (2006): 42–71.

Lovink, Geert. *Networks without a Cause: A Critique of Social Media.* Cambridge, UK: Polity, 2012.

Lyman, Peter, and Hal R. Varian. "How Much Information?" *How Much Information? 2003,* 2003. Available at http://www2.sims.berkeley.edu/re search/projects/how-much-info-2003/.

Mackenzie, Adrian. *Cutting Code: Software and Sociality.* New York: Peter Lang, 2006.

———. "The Mortality of the Virtual: Real-Time, Archive and Dead-Time in Information Networks." *Convergence* 3, no. 2 (1997).

MacManus, Richard. "Interview with Digg Founder Kevin Rose, Part 1." *ZDNet,* February 1, 2006. Available at http://blogs.zdnet.com/web2explorer/index .php?p=108.

Mahdian, Alireza, John Black, Richard Han, and Shivakant Mishra. "MyZone: A Next-Generation Online Social Network." *arXiv:1110.5371* (October 24, 2011). Available at http://arxiv.org/abs/1110.5371.

Mahoney, Michael Scan. *Histories of Computing.* Edited by Thomas Haigh. Cambridge, MA: Harvard University Press, 2011.

Manoff, Marlene. "Theories of the Archive from across the Disciplines." *Libraries and the Academy* 4 (2004): 9–25.

Markley, Robert. "Boundaries: Mathematics, Alienation, and the Metaphysics of Cyberspace." *Configurations* 2, no. 3 (1994): 485–507. doi:10.1353/con .1994.0037.

Martinez-Torres, M. E. "Civil Society, the Internet, and the Zapatistas." *Peace Review* 13, no. 3 (2001): 347–355.

Marwick, Alice. "Status Update: Celebrity, Publicity and Self-Branding in Web 2.0." Dissertation, New York University, 2010.

———. "To Catch a Predator? The MySpace Moral Panic." *First Monday* 13, no. 6 (June 2, 2008). Available at http://firstmonday.org/htbin/cgiwrap/bin/ojs/ index.php/fm/article/view/2152/1966.

Marwick, Alice, and danah boyd. "I Tweet Honestly, I Tweet Passionately: Twitter Users, Context Collapse, and the Imagined Audience." *New Media and Society* (July 2010). doi:10.1177/1461444810365313.

———. "To See and Be Seen: Celebrity Practice on Twitter." *Convergence: The International Journal of Research into New Media Technologies* 17, no. 2 (May 1, 2011): 139–158. doi:10.1177/1354856510394539.

Marx, Karl. *A Contribution to the Critique of Political Economy.* Marxists.org, 1859. Available at http://www.marxists.org/archive/marx/works/1859/critique-pol-economy/preface.htm.

———. "Economic and Philosophical Manuscripts of 1844." *Marxists Internet Archive,* 2009. Available at http://www.marxists.org/archive/marx/works/ 1844/manuscripts/preface.htm.

———. *Grundrisse: Foundations of the Critique of Political Economy (Rough Draft).* London: Penguin Books in association with New Left Review, 1993.

———. *The Poverty of Philosophy: Answer to* The Philosophy of Poverty *by M. Proudhon,* 1847. Available at http://www.marxists.org/archive/marx/ works/1847/poverty-philosophy/index.htm.

———. "Theses on Feuerbach." Archive. *Marxists Internet Archive,* 2002. Available at http://www.marxists.org/archive/marx/works/1845/theses/theses.htm.

Marx, Karl, and Martin Nicolaus. *Grundrisse.* New York: Penguin Classics, 1993.

Mattelart, Armand. *Networking the World, 1794–2000.* English language. Minneapolis: University of Minnesota Press, 2000.

Mattus, C. J. "Legal Update: Is It Really My Space? Public Schools and Student Speech on the Internet After Layshock V. Hermitage School District and Snyder V. Blue Mountain School District." *Boston University Journal of Science and Technology Law* 16 (2010): 318.

Mauldin, M. L. "Chatterbots, Tinymuds, and the Turing Test: Entering the Loebner Prize Competition." In *Proceedings of the National Conference on Artificial Intelligence*, 16–21, 1994.

Maxwell, Richard, and Toby Miller. *Greening the Media*. Oxford, UK: Oxford University Press, 2012.

McChesney, Robert W. "Conflict, Not Consensus: The Debate over Broadcast Communication Policy." In *Ruthless Criticism: New Perspectives in U.S. Communications History*, 222–258. Minneapolis: University of Minnesota Press, 1993.

———. *The Political Economy of Media: Enduring Issues, Emerging Dilemmas*. New York: Monthly Review Press, 2008.

Mediati, Nick. "Windows 7: How We Test." *PC World*, September 29, 2009. Available at http://www.pcworld.com/article/172510/windows_7_how_we_test.html.

———. "Windows 7 Performance Tests." *PC World*, October 19, 2009. Available at http://www.pcworld.com/article/172509/windows_7_performance_tests.html.

Messer-Kruse, Timothy. "The 'Undue Weight' of Truth on Wikipedia." *Chronicle of Higher Education*, February 12, 2012, sec. The Chronicle Review. Available at http://chronicle.com/article/The-Undue-Weight-of-Truth-on/130704/.

Mika, Peter. *Social Networks and the Semantic Web*. New York: Springer, 2007.

Montfort, Nick, and Ian Bogost. *Racing the Beam: The Atari Video Computer System*. Cambridge, MA: MIT Press, 2009.

Moody, Fred. *I Sing the Body Electronic: A Year with Microsoft on the Multimedia Frontier*. New York: Penguin Books, 1995.

Moody, Glyn, and Eben Moglen. "Interview: Eben Moglen—Freedom vs. the Cloud Log." *The H*, March 17, 2010. Available at http://www.h-online.com/open/features/Interview-Eben-Moglen-Freedom-vs-the-Cloud-Log-955421.html.

Morozov, Evgeny. "Muzzled by the Bots." *Slate*, October 26, 2012. Available at http://www.slate.com/articles/technology/future_tense/2012/10/disintermediation_we_aren_t_seeing_fewer_gatekeepers_we_re_seeing_more.html.

Morrissey, Brian. "MySpace Is Still Here." *Brandweek*, August 30, 2010. Available at http://www.brandweek.com/bw/content_display/esearch/e3i84260d4301c885f95e2bd7d033dc1541?pn=1.

Mosco, Vincent. *The Digital Sublime: Myth, Power, and Cyberspace*. Cambridge, MA: MIT Press, 2004.

Mueller, Gavin. "Piracy Is Looting—And That's OK." *Unfashionably Late*, August 29, 2011. Available at http://unfashionablylate.wordpress.com/2011/08/29/piracy-is-looting-and-thats-ok/.

Mueller, M. "ICANN and Internet Governance: Sorting through the Debris of 'Self-Regulation.'" *Info* 1, no. 6 (1999): 497–520.

Nadesan, Majia Holmer. "Transparency and Neoliberal Logics of Corporate Economic and Social Responsibility." In *The Handbook of Communication and Corporate Social Responsibility*, edited by Øyvind Ihlen, Jennifer L. Bartlett, and Steve May, 252–275. Malden, MA: Wiley-Blackwell, 2011. Available at http://onlinelibrary.wiley.com/doi/10.1002/9781118083246.ch13/summary.

Nandhakumar, Joe. "Managing Time in a Software Factory: Temporal and Spatial Organization of IS Development Activities." *Information Society* 18, no. 4 (July 2002): 251–262. doi:10.1080/01972240290075101.

Nanis, Max, Ian Pearce, and Tim Hwang. "PacSocial: Field Test Report." The Pacific Social Architecting Corporation, November 15, 2011.

Naur, P., and B. Randell. "Software Engineering: Report of a Conference Sponsored by the NATO Science Committee, Garmisch, Germany, 7th to 11th October, 1968," 1969.

Naylor, M., and C. Runciman. "The Reduceron: Widening the von Neumann Bottleneck for Graph Reduction Using an FPGA." *Selected Papers from the Proceedings of IFL 2007* (2008).

Neumann, Peter G. "Inside Risks: Linguistic Risks." *Commun. ACM* 39, no. 5 (May 1996): 154. doi:10.1145/229459.247239.

Noble, David F. *America by Design: Science, Technology, and the Rise of Corporate Capitalism*. New York: Knopf, 1977.

O'Malley, Gavin. "MySpace Blossoms into Major Web Portal." *Advertising Age* 77, no. 29 (July 17, 2006): 4–26.

Optimize IT. "Facebook Wants to Issue Your Internet Driver's License." *Facebook*, January 16, 2011. Available at http://www.facebook.com/note.php?note_id=500459938286.

O'Reilly, Tim. "The Architecture of Participation." *O'Reilly Media*, June 2004. Available at http://oreilly.com/pub/a/oreilly/tim/articles/architecture_of_participation.html.

———. "What Is Web 2.0: Design Patterns and Business Models for the Next Generation of Software." *Communications and Strategies* 65 (2007): 17–37.

O'Reilly, Tim, Andrew Anker, Brian Behlendorf, Alan Vermeulen, and Bog Morgan. "The Architecture of Participation." Web 2.0 Conference, San Francisco, 2004.

O'Reilly, Tim, and John Battelle. "Opening Welcome: State of the Internet Industry." Web 2.0 Conference, San Francisco, 2004.

Ostrow, Adam. "Digg Bans Company That Blatantly Sells Diggs." *Mashable*, 2009. Available at http://mashable.com/2008/12/02/digg-bans-company-that-blatantly-sells-diggs/.

Pagendarm, M., and H. Schaumburg. "Why Are Users Banner-Blind? The Impact of Navigation Style on the Perception of Web Banners." *Journal of Digital Information* 2, no. 1 (2006). Available at http://journals.tdl.org/jodi/index.php/jodi/article/view/36/38.

Paolillo, J. C., and E. Wright. "Social Network Analysis on the Semantic Web: Techniques and Challenges for Visualizing FOAF." *Visualizing the Semantic Web: XML-Based Internet and Information Visualization* (2006): 229–241.

Parnas, D. L., and D. P. Siewiorek. "Use of the Concept of Transparency in the

Design of Hierarchically Structured Systems." *Communications of the ACM* 18, no. 7 (July 1975): 401–408. doi:10.1145/360881.360913.

Perelman, Michael. *The Perverse Economy: The Impact of Markets on People and the Environment.* New York: Palgrave Macmillan, 2003.

Perez, Juan Carlos. "MySpace Slashes Nearly Half of Staff." *PC World,* January 11, 2011. Available at http://www.pcworld.com/businesscenter/article/216468/ myspace_slashes_nearly_half_of_staff.html.

Phillips, Sarah. "A Brief History of Facebook." *Guardian,* July 25, 2007. Available at http://www.guardian.co.uk/technology/2007/jul/25/media.newmedia.

"Por Qué Estamos Aquí y No en Es.wikipedia.org." *La Enciclopedia Libre Universal en Español,* July 25, 2007. Available at http://enciclopedia.us.es/ index.php/Enciclopedia:Por_qu%C3%A9_estamos_aqu%C3%AD_y_no_ en_es.wikipedia.org.

Postigo, Hector. "America Online Volunteers: Lessons from an Early Coproduction Community." *International Journal of Cultural Studies* 12, no. 5 (September 1, 2009): 451–469. doi:10.1177/1367877909337858.

———. "Emerging Sources of Labor on the Internet: The Case of America Online Volunteers." *International Review of Social History* 48, no. S11 (December 2003): 205–223. doi:10.1017/S0020859003001329.

———. "Of Mods and Modders: Chasing Down the Value of Fan-Based Digital Game Modifications." *Games and Culture* 2, no. 4 (October 2007): 300–313. doi:10.1177/1555412007307955.

Poulsen, Kevin. "MySpace Predator Caught by Code." Magazine. *Wired,* October 16, 2006. Available at http://www.wired.com/science/discoveries/news/ 2006/10/71948.

PricewaterhouseCoopers. *IAB Online Ad Measurement Study.* Interactive Advertising Bureau, December 2001.

"PricewaterhouseCoopers: Global: Insights and Solutions: MoneyTree™ Survey Report," 2009. Available at https://www.pwcmoneytree.com/MTPublic/ns/ nav.jsp?page=historical.

Puckett, J., L. Byster, S. Westervelt, R. Gutierrez, S. Davis, A. Hussain, and M. Dutta. "Exporting Harm: The High-Tech Trashing of Asia." Seattle, WA: Basel Action Network and Silicon Valley Toxics Coalition. Available at http:// www.Ban.org/E-waste/technotrashfinalcomp.Pdf (2003).

Rancier, J. "The Concept of 'Critique' in the 'Critique of Political Economy.'" In *Ideology, Method, and Marx,* edited by A. Rattansi. London: Routledge, 1989.

Raphael, J. R. "Myspace's Facebook 'Mashup'—Why Bother?" *PC World,* November 18, 2010. Available at http://www.pcworld.com/article/211127/ myspaces_facebook_mashup_why_bother.html#tk.mod_rel.

Raymond, E. S. *The Cathedral and the Bazaar: Musings on Linux and Open Source by an Accidental Revolutionary.* Sebastopol, CA: O'Reilly and Associates, 2001.

Rimmer, M. "Napster: Infinite Digital Jukebox or Pirate Bazaar?" 2001. Available at http://works.bepress.com/matthew_rimmer/38.

Risen, James, and Eric Lichtblau. "Bush Lets U.S. Spy on Callers without Courts." *New York Times,* December 16, 2005, sec. Washington. Available at https:// www.nytimes.com/2005/12/16/politics/16program.html.

Robins, K., and F. Webster. "'The Revolution of the Fixed Wheel': Information, Technology and Social Taylorism." *Television in Transition*. London: BFI (1985): 36–63.

Rosen, Rebecca J. "Does Wikipedia Have an Accuracy Problem?" *The Atlantic,* February 16, 2012. Available at http://www.theatlantic.com/technology/ar chive/2012/02/does-wikipedia-have-an-accuracy problem/253216/.

Rosenberg, Scott. *Dreaming in Code: Two Dozen Programmers, Three Years, 4,732 Bugs, and One Quest for Transcendent Software.* New York: Crown Publishers, 2007.

Ruiz Rivas, Juanan Antonio. "Usuario: Juanan." *La Enciclopedia Libre Universal en Español,* February 2002. Available at http://enciclopedia.us.es/index.php/ Usuario:Juanan.

Russell, A. L. "'Industrial Legislatures': Consensus Standardization in the Second and Third Industrial Revolutions." *Enterprise and Society* 10, no. 4 (2009): 661.

Salter, L. "Democracy, New Social Movements and the Internet." *Cyberactivism: Online Activism in Theory and Practice* (2003): 117–144.

Sanger, Larry. "Announcement about My Involvement in Wikipedia and Nupe-dia." *OSDir,* February 13, 2002. Available at http://osdir.com/ml/science.lin guistics.wikipedia.international/2002-02/msg00037.html.

———. "Re: [Intlwiki-l]." *OSDir,* February 19, 2002. Available at http://osdir .com/ml/science.linguistics.wikipedia.international/2002-02/msg00058 .html.

———. "Why Wikipedia Must Jettison Its Anti-Elitism." *Kuro5hin,* December 31, 2004. Available at https://www.kuro5hin.org/story/2004/12/30/142458/25.

Schaefer, Peter D., and Meenakshi Gigi Durham. "On the Social Implications of Invisibility: The iMac G5 and the Effacement of the Technological Object." *Critical Studies in Media Communication* 24, no. 1 (March 2007): 39–56. doi:10.1080/07393180701214520.

Scharman, Fred. "'You Must Be Logged in to Do That!': Myspace and Control." *SevenSixFive,* May 2006. Available at http://www.sevensixfive.net/myspace/ myspacetwopointoh.html.

Schiller, Dan. *How to Think about Information.* Urbana: University of Illinois Press, 2007.

Scholz, Trebor. "Market Ideology and the Myths of Web 2.0." *First Monday* 13, no. 3 (March 3, 2008). Available at http://firstmonday.org/article/view/2138/ 1945.

Seife, Charles. *Decoding the Universe: How the New Science of Information Is Explaining Everything in the Cosmos, from Our Brains to Black Holes.* New York: Penguin Books, 2007.

Sekula, A. "Reading an Archive: Photography between Labour and Capital." In *Visual Culture: The Reader,* edited by Jessica Evans and Stuart Hall, 181–192. London, UK: SAGE Publications, 1999.

Sengupta, Somini. "So Much for Sharing His 'Like.'" *New York Times,* May 31, 2012, sec. Technology. Available at https://www.nytimes.com/2012/06/01/ technology/so-much-for-sharing-his-like.html.

Shannon, C. E. "The Mathematical Theory of Communication." In *The Mathematical Theory of Communication*, 3–93. Urbana-Champaign: University of Illinois Press, 1949.

Shapiro, Ari. "Obama, Romney Campaigns Taking 'See What Sticks' Approach To Web Videos." *NPR.org*, June 11, 2012. Available at http://www.npr.org/blogs/itsallpolitics/2012/06/11/154604039/obama-romney-campaigns-taking-see-what-sticks-approach-to-web-videos.

Shields, Mike. "Facebook Trumps MySpace on Ads." *Brandweek*, August 13, 2010. Available at http://www.brandweek.com/bw/content_display/esearch/e3i38fc3a9296f214d3cfcafd70a5a89722.

Shirky, Clay. "It's Not Information Overload. It's Filter Failure." Conference presentation at the Web 2.0 Expo, San Francisco, April 23, 2008. Available at http://web2expo.blip.tv/file/1277460.

Singer, Natasha. "Acxiom, the Quiet Giant of Consumer Database Marketing." *New York Times*, June 16, 2012, sec. Technology. Available at https://www.nytimes.com/2012/06/17/technology/acxiom-the-quiet-giant-of-consumer-database-marketing.html.

Smith, Adam. *The Wealth of Nations*. New York: Bantam Classics, 2003.

Smith, Paul. *Discerning the Subject*. Minneapolis: University of Minnesota Press, 1988.

Social Network Websites: Best Practices from Leading Services. FaberNovel Consulting, November 28, 2007. Available at http://www.fabernovel.com/socialnetworks.pdf.

Sohn-Rethel, Alfred. *Intellectual and Manual Labour: A Critique of Epistemology*. London, UK: Macmillan Press, 1983.

Sparrow, Elijah. "Pitfalls of Building Social Media Alternatives (Debate)." Presented at Unlike Us #2, Amsterdam, March 10, 2012. Available at http://vimeo.com/39257353.

———. "Social Media and the Rise of Surveillance-Based Advertising." Presented at Cyber-Surveillance in Everyday Life, Toronto, 2011.

Spideralex. "Pitfalls of Building Social Media Alternatives (Debate)." Presented at Unlike Us #2, Amsterdam, March 10, 2012. Available at http://vimeo.com/39257151.

Spolsky, Joel. "The Law of Leaky Abstractions." *Joel on Software*, November 11, 2002. Available at http://www.joelonsoftware.com/articles/LeakyAbstractions.html.

Star, Susan Leigh. "The Ethnography of Infrastructure." *American Behavioral Scientist* 43, no. 3 (1999): 377–391.

Stross, Randall E. *The Microsoft Way: The Real Story of How the Company Outsmarts Its Competition*. Reading, MA: Addison-Wesley, 1996.

Sullivan, Mark. "Myspace to Use Facebookers' Entertainment Likes." *PC World*, November 18, 2010. Available at http://www.pcworld.com/article/211095/myspace_to_use_facebookers_entertainment_likes.html?tk=hp_new.

Swann, G.M.P. "The Economics of Standardization." *Final Report for the Standards and Technical Regulations Directorate, Department of Trade and Industry*. Manchester Business School, University of Manchester, 2000.

Szattler, Eduard. "GPL Viral License or Viral Contract." *Masaryk University Journal of Law and Technology* 1 (2007): 67.

Szoka, B. M., and A. D. Thierer. "Cyberbullying Legislation: Why Education Is Preferable to Regulation." *Progress and Freedom Foundation Progress on Point Paper* 16, no. 12 (June 2009).

Szpir, Michael. "Clickworkers on Mars." *American Scientist,* June 2002. Available at http://www.americanscientist.org/issues/pub/clickworkers-on-mars.

Taylor, C. *The Culture of Confession from Augustine to Foucault: A Genealogy of the "Confessing Animal."* New York: Taylor and Francis, 2008.

Terranova, Tiziana. "Free Labor: Producing Culture for the Digital Economy." *Social Text* 18, no. 2 (2000): 33–58.

————. *Network Culture: Politics for the Information Age.* London: Pluto Press, 2004.

"30 Jan Stats." *OSDir,* January 30, 2002. Available at http://osdir.com/ml/science.linguistics.wikipedia.international/2002-01/msg00120.html.

Tkacz, Nathaniel. "The Politics of Forking Paths." In *Critical Point of View a Wikipedia Reader,* edited by Geert Lovink and Nathaniel Tkacz, 94–107. Amsterdam: Institute of Network Cultures, 2011.

Toscano, Alberto. "The Open Secret of Real Abstraction." *Rethinking Marxism* 20, no. 2 (2008): 273–273.

Townes, Miles. "The Spread of TCP/IP and the Political Origins of the Internet." Dissertation, University of Maryland, College Park, 2011.

Turbert, K. "Faceless Bullies: Legislative and Judicial Responses to Cyberbullying." *Seton Hall Legislative Journal* 33 (2009): 651.

Turing, Alan M. "Computing Machinery and Intelligence." *Mind* 59, no. 236 (1950): 433–460.

————. *The Essential Turing: Seminal Writings in Computing, Logic, Philosophy, Artificial Intelligence, and Artificial Life Plus The Secrets of Enigma.* Edited by B. Jack Copeland. Oxford, UK: Oxford University Press, 2004.

Turkle, Sherry. "Can You Hear Me Now?" *Forbes,* May 7, 2007. Available at http://www.forbes.com/free_forbes/2007/0507/176.html.

————. *Life on the Screen: Identity in the Age of the Internet.* New York: Simon and Schuster, 1995.

Turner, Fred. *From Counterculture to Cyberculture: Stewart Brand, the Whole Earth Network, and the Rise of Digital Utopianism.* Chicago: University of Chicago Press, 2008.

Tynan, Dan. "The 25 Worst Web Sites." *PC World,* September 15, 2006. Available at http://www.pcworld.com/article/127116-7/the_25_worst_web_sites.html.

Vance, Ashlee. "This Tech Bubble Is Different." *BusinessWeek: Online Magazine,* April 14, 2011. Available at http://www.businessweek.com/magazine/content/11_17/b4225060960537.htm.

Vega, Tanzina, and Verne Kopytoff. "Web Advertisers Fear Effects of Do-Not-Track System." *New York Times,* December 5, 2010, sec. Business Day/Media and Advertising. Available at http://www.nytimes.com/2010/12/06/business/media/06privacy.html?_r=1.

Virilio, P. *The Art of the Motor.* Minneapolis: University of Minnesota Press, 1995.

————. *Negative Horizon.* Translated by Michael Degener. New York: Continuum, 2005.

———. *Speed and Politics: An Essay on Dromology*. Translated by Mark Politzzotti. New York: Semiotext(e), 1986.

Virtanen, Akseli. "General Economy: The Entrance of Multitude into Production." *Ephemera: Theory and Politics in Organization* 4, no. 3 (2004): 209–232.

Vollmer, Christopher. *Digital Darwinism*. Alrington, VA: Booz, 2009. www.booz.com/media/uploads/Digital_Darwinism.pdf.

Von Neumann, John. "First Draft of a Report on the EDVAC." *IEEE Annals of the History of Computing* 15, no. 4 (1993): 27–43.

Warf, B., and J. Grimes. "Counterhegemonic Discourses and the Internet." *Geographical Review* 87, no. 2 (1997): 259–274.

Wark, McKenzie. *A Hacker Manifesto*. Cambridge, MA: Harvard University Press, 2004.

Webb, Judson Chambers. *Mechanism, Mentalism, and Metamathematics: An Essay of Finitism*. Dordrecht, Netherlands: D. Reidel, 1980.

Weber, Steven. *The Success of Open Source*. Cambridge, MA: Harvard University Press, 2004.

Webster, Stephen C. "Military's 'Persona' Software Cost Millions, Used for 'Classified Social Media Activities.'" *The Raw Story*, February 22, 2011. Available at http://www.rawstory.com/rs/2011/02/22/exclusive-militarys-persona-software-cost-millions-used-for-classified-social-media-activities/.

Wegrzanowski, Tomasz. "Polish Wikipedia and PHP." *OSDir*, March 15, 2002. Available at http://osdir.com/ml/science.linguistics.wikipedia.international/2002-03/msg00051.html.

Weiss, M., and C. Cargill. "Consortia in the Standards Development Process." *Journal of the American Society for Information Science* 43, no. 8 (1992): 559–565.

Weizenbaum, Joseph. *Computer Power and Human Reason: From Judgment to Calculation*. San Francisco: W. H. Freeman, 1976.

Wells, William D., and Qemei Chen. "The Dimensions of Commercial Cyberspace." *Journal of Interactive Advertising* 1, no. 1 (Fall 2000).

Whitaker, J. L, and B. J Bushman. "Online Dangers: Keeping Children and Adolescents Safe." *Washington and Lee Law Review* 66 (2009): 1053.

Whitehead, Alfred North. *Science and the Modern World*. New York: Free Press, 1997.

Wiener, N. *Cybernetics: Or, Control and Communication in the Animal and the Machine*. Cambridge, MA: MIT Press, 1965.

———. *The Human Use of Human Beings: Cybernetics and Society*. London, UK: Da Capo Press, 1988.

Wiese, Karen. "On Diaspora's Social Network, You Own Your Data." *Businessweek.com*, May 10, 2012. Available at http://www.businessweek.com/printer/articles/24762-on-diasporas-social-network-you-own-your-data.

Willard, N. "Cyberbullying Legislation and School Policies: Where Are the Boundaries of the 'Schoolhouse Gate' in the New Virtual World?" *Retried June* 12 (2007): 2008.

Williamson, Debra Aho. "Groups Set to Unveil Web Ad Guidelines." *Advertising Age* 67, no. 50 (December 9, 1996): 1.

————. "Interactive: Site Comparability Top Priority in IAB Guidelines." *Advertising Age,* July 16, 1997: 48.

————. "Internet Ad Bureau to Tally Ad Spending." *Advertising Age* 67, no. 26 (June 24, 1996): 4.

Winner, Langdon. *The Whale and the Reactor: A Search for Limits in an Age of High Technology.* Chicago: University of Chicago Press, 1986.

Woolgar, Steve. "Configuring the User: The Case of Usability Trials." In *A Sociology of Monsters: Essays on Power, Technology and Domination,* edited by John Law, 57–99. London: Routledge, 1991.

Wray, Richard. "Internet Data Heads for 500bn Gigabytes." *The Guardian,* May 18, 2009. Available at http://www.guardian.co.uk/business/2009/may/18/digital-content-expansion.

Wright, E. O., ed. "Foundations of a Neo-Marxist Class Analysis." In *Approaches to Class Analysis,* 1–26. Cambridge, UK: Cambridge University Press, 2005. Available at http://www.ssc.wisc.edu/~wright/Chapter%201%20--%20Wright%20Jan%202004.pdf.

————. "A General Framework for the Analysis of Class Structure." In *Class Analysis,* 64–104. London, UK: Verso, 1977.

Yadav, Sid. "Facebook—The Complete Biography." *Mashable,* August 25, 2006. Available at http://mashable.com/2006/08/25/facebook-profile/.

Zittrain, Jonathan. *The Future of the Internet and How to Stop It.* New Haven, CT: Yale University Press, 2008.

————. "Minds for Sale." *Berkman Center for Internet and Society,* November 18, 2009. Available at http://cyber.law.harvard.edu/interactive/events/2009/11/berkwest.

Zuckerberg, Mark. "A Conversation with Mark Zuckerberg." Interview presented at the Web 2.0 Summit, San Francisco, November 15, 2010. Available at http://itc.conversationsnetwork.org/shows/detail4757.html.

Zuckerman, Lawrence. "Firms Competing to Define Web Demographics." *Stuart News.* April 27, 1997, Martin County edition.

Zwick, D., S. K. Bonsu, and A. Darmody. "Putting Consumers to Work: Co-creation and New Marketing Govern-Mentality." *Journal of Consumer Culture* 8, no. 2 (2008): 163–196.

Index

Abstraction: conceptual, 10, 18, 90, 109–110, 169n47; real, 17–18, 72–73, 76–82, 87, 109, 135–136; software engineering, use in, 10, 17, 58, 65, 72–76, 87–90, 156, 161

ACTA (Anti-counterfeiting Trade Agreement), 151

Activism: Arab Spring, 6, 142, 164; class-based, 124–126, 162; media, 15, 20, 116, 128, 143–144, 157–158, 160, 162–165, 189n38; Occupy Wall Street, 6, 142, 164

Actor-network theory, 12. *See also* Association; Heterogeneous engineering

Advertising: ad-blocking software, 114–115, 178n45; advertising networks, 35, 71, 102, 104, 112, 115, 136; eye-tracking studies supporting, 100, 105; resistance to, 84, 108–109, 114, 119, 131, 133–134, 178n45; and social media, 56–57, 63, 68, 84–87, 88, 90, 105–108, 110–111, 174n58, 186n21; standardization of, 93, 98–104, 109, 126–127, 181n42, 181–182n48, 183–184n76

Affect: affective processing, 42–43, 65, 67, 125; archives of, 43, 62, 65, 68, 107–108; labor and, 52, 82, 88, 118; in social media, 49, 51, 72–73, 123, 155

Agency, 12–14, 20, 95, 99, 118, 120

Aggregation: of data sets, 30, 34, 66, 174n49; of servers, 145–146; by socialbots, 21; of user labor, 35, 42, 53, 57, 90, 113, 116, 123, 174n41

AJAX (Asynchronous Javascript and XML), 14, 41, 50, 58, 61–62, 169n49, 175n63

Alienation, 23, 121, 152, 182n60

Amazon, 47, 98, 103, 145; Mechanical Turk, 43, 57–59, 61; user-generated reviews in, 2, 5

America Online (AOL), 98, 101, 104, 127

Andrejevic, Mark, 67, 81, 110, 117

Anonymity, 55, 137, 155, 158, 160–162, 179n72

Anti-counterfeiting Trade Agreement (ACTA), 151

AOL (America Online), 98, 101, 104, 127

API (Application Programming Interface), 18, 63, 93, 113, 184n78

Robert W. Gehl is an Assistant Professor in the Department of Communication at the University of Utah. He is co-editor (with Victoria Watts) of *The Politics of Cultural Programming in Public Spaces*.